Mel Turner barn, Sullivan Township. (photograph by Pief L. Weyman)

Barns of the Queen's Bush

Jon Radojkovic

with drawings by Mary Tripp MacCarl

The Brucedale Press

ISBN 1-896922-16-3 (bound) — ISBN 1-896922-15-5 (pbk.)
The Brucedale Press
Box 2259
Port Elgin, Ontario N0H 2C0

Unless credited to an archival source or shown with a lender's name, all photographs are by the author, and copyright rests with him.

While care has been taken to ensure accuracy and completeness of the text, the author welcomes any further information on any aspect of the book.

While the author and publisher encourage appreciation for the barns in the Queen's Bush landscape and elsewhere, we remind travellers to respect the privacy of barn owners, and always obtain permission to enter private property.

National Library of Canada Cataloguing in Publication Data

Radojkovic, Jon, 1952-
Barns of the Queen's bush

ISBN 1-896922-16-3 (bound) — ISBN 1-896922-15-5 (pbk.)

1. Barns—Ontario—Grey (County)—History. 2. Barns—Ontario—Bruce (County)—History. 3. Barns—Ontario—Grey (County)—Pictorial works. 4. Barns—Ontario—Bruce (County)—Pictorial works. 5. Grey (Ont. : County)—History, Local. 6. Bruce (Ont. : County)—History, Local. I. Title.

NA8230.R33 2001 631.2'2'0971318 C2001-902125-9

Edited for publication by Anne Judd.
Design, prepress, and printing by St. Jacobs Printery Ltd., St. Jacobs, Ontario.
Created and produced entirely in Canada.

COVER IMAGES — FRONT LEFT: *Barn raising, 1910, Holland Township. (courtesy of Grey County-Owen Sound Museum);* ***FRONT RIGHT:*** *Barn ablaze, 1981. (courtesy of Glen Stephens);* ***BACK:*** *Threshing scene, circa 1905, Euphrasia Township. (courtesy of Glen Stephens);* ***BACKGROUND:*** *Timber details, the biggest barn in Grey County. (photo by the author)*

dedicated to all the pioneers who built these great timber-frame structures

(courtesy Bruce County Museum & Archives)

Author's Acknowledgements

I would like to acknowledge the assistance of my wife, Lillian Burgess, with special thanks for her input and advice, and of my whole family, who put up with my lack of time for anything else while I was writing. My gratitude to Mary Tripp MacCarl, who took my ideas and drew them into illustrations. Thanks to the staff of the Chesley Centennial Library, especially Mary Witzke, for being so helpful. My appreciation to the staff of the County of Grey-Owen Sound Museum and Bruce County Museum & Archives for their assistance. To my friends Pief Weyman and Dinand van der Wal, I offer thanks for their photo darkroom expertise. A big thanks to Anne Duke Judd whose belief in me and gentle prodding finally moved me to write this book. To all the wonderful people I interviewed, and those who loaned me not only pictures, but shared a piece of their lives—thank-you.

Contents

Introduction

My adventure with timber framing began in 1984 when I moved to Sullivan Township in Grey County with my wife, Lillian, and our newly born son, Skyler. We had purchased 96 acres [39 hectares] of fields, swamp, and hardwood bush. No house, barn, shed, or any other structure adorned this land but as I looked at our forest I saw material readily available for all our needs. Although I was a city kid, I had acquired some carpentry skills from my father. We had no money to hire anyone and nothing to lose—but something to gain—if I tried building myself. We decided to build a 1,600-square foot [150 square metre] timber-frame home, using wood from the bush.

I began by learning about the different types of trees in our 30 acre [12 hectare] forest, all of the species prevalent in Grey and Bruce Counties: maple, white ash, beech, hemlock, black cherry, butternut, aspen, ironwood, cedar and white pine. How fortunate I was to stumble upon such a prize! Of course, in 1984, I didn't know how lucky I was—I thought all forests in Ontario were the same. It was later, when I began to discover more about Ontario's tree cover, that I realized the value of our bush.

After cutting a clearing in the forest big enough for a house site and building a workshop—which we lived in for four years—I began selecting the trees to saw into beams for our house. At that time a small sawmill on the concession north of us was operated by Noah Shettler, a smart, friendly Amish man whom I got to know very well. All during the winter and spring of 1985 I cut down the trees and then pulled them out with our 1952 Farmall Super M. Since I did not have a front-end loader, I rigged a block and tackle to a tree, winched the huge hardwood logs onto the wagon, and then took them to the sawmill in twos and threes.

In the meantime we planted some grain, cut hay for cash and began the first of our garlic crops. We learned quickly that farming is tough. Even though Lillian had practical experience and a degree in agriculture, we soon realized that it would be a difficult way to make a living, especially since our neighbours, who had been farming all their lives, were having a hard time of it.

Once I began bringing the sawn beams from the mill, it was time for notching. All I can say about notching hardwood beams is, "It's a good thing I didn't know any better." There was no electrical power available on our concession road and we decided it would be too expensive to bring it the half-mile from the nearest source. That meant no electric drills or saws. Instead, I used a chain saw for cutting the beams to the correct length, a fifty-year-old boring machine for mortises, and some chisels I had acquired at auction sales for finishing the notches. Using a boring machine can be a meditative experience. With a two-inch [5 cm] bit on the end and a white ash or maple beam staring you in the face, you can become quite reflective about your inner strengths, not to mention your outer ones. I set up a routine, notching in the mornings and doing the hundred other things that needed to be done in the afternoons and evenings. I also took down a barn, with a neighbour's help, and incorporated some of those beams into the house structure.

I had taken a three-week timber-frame course at the Shelter Institute in Maine, USA, and a week-end workshop at Thistlewood Timber Frames in Markdale, Ontario. Although both courses held forth on the attributes of big electric saws and drills, I did learn some of the basics. Mostly I notched by the seat of my pants. I began to appreciate what the early farmers in the Queen's Bush accomplished and how much work that took.

In late October of 1987, my bents were ready to be raised. Each of the four bents was 18 feet wide and 18 feet high, [5.5 metres x 5.5 metres] with a crossbeam in the middle for the second floor. An extension 10 feet high and 10 feet wide [3 metres x 3 metres] completed the saltbox shape with a high south and low north side. I used a variety of hardwood beams, choosing whatever was straight and large enough to cut from our bush.

The morning of the raising dawned grey and bitterly cold with rain and sleet. Our new neighbours and old friends came out to help anyway and, with the aid of a crane, we raised our two-storey home. What a momentous thrill it was for me to see those hardwood bents come together and fit. Since then I have notched more house and barn frames, and the thrill of raising day still amazes me. That a few beams, notched precisely, and strengthened with braces, can make a structure to last many lifetimes, seems incredible.

After our raising, I began to notice and admire the barns in our area. Our house was small compared to these huge structures, and they were constructed with even fewer power tools than I had.

I also began to realize that many barns were looking neglected. Some were being pulled down for the material or destroyed by storms. They would never be rebuilt. The urge grew in me to record the history of these amazing buildings, as well as the stories of the people who made and lived with them. The way I set about doing that was to interview present and past owners and to photograph the barns.

Everyone was welcoming and allowed me to roam within their barns or mills and farmyards just as I pleased. Interviews invariably led to the kitchen table, and offers of coffee, lunch and "Call back if you need more information." In all my many conversations with these kind folk there was one prevailing comment: "We couldn't have done it without the help of the community." Not only for barn raisings but also threshing bees, haying, beef rings, and sharing of farm implements, community life was almost like a marriage. "In sickness and in health" neighbours were always there to help. That's how it used to be, but that way of life is mostly gone. What I hope to record in this book is a bit of the social history of the Queen's Bush that revolved around the farm and its centrepiece—the barn.

Jon Radojkovic (aka Jon Log)
June 2001

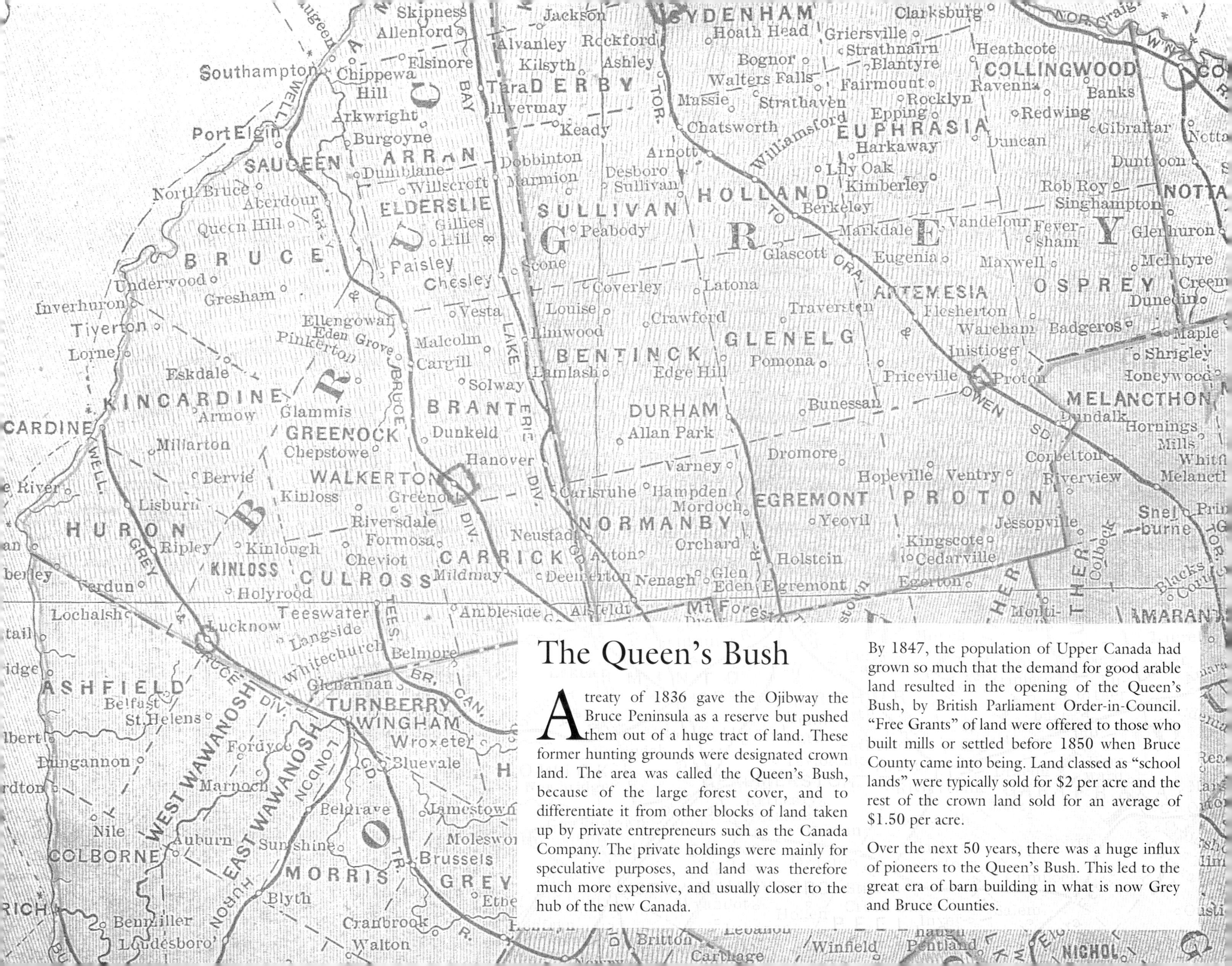

The Queen's Bush

A treaty of 1836 gave the Ojibway the Bruce Peninsula as a reserve but pushed them out of a huge tract of land. These former hunting grounds were designated crown land. The area was called the Queen's Bush, because of the large forest cover, and to differentiate it from other blocks of land taken up by private entrepreneurs such as the Canada Company. The private holdings were mainly for speculative purposes, and land was therefore much more expensive, and usually closer to the hub of the new Canada.

By 1847, the population of Upper Canada had grown so much that the demand for good arable land resulted in the opening of the Queen's Bush, by British Parliament Order-in-Council. "Free Grants" of land were offered to those who built mills or settled before 1850 when Bruce County came into being. Land classed as "school lands" were typically sold for $2 per acre and the rest of the crown land sold for an average of $1.50 per acre.

Over the next 50 years, there was a huge influx of pioneers to the Queen's Bush. This led to the great era of barn building in what is now Grey and Bruce Counties.

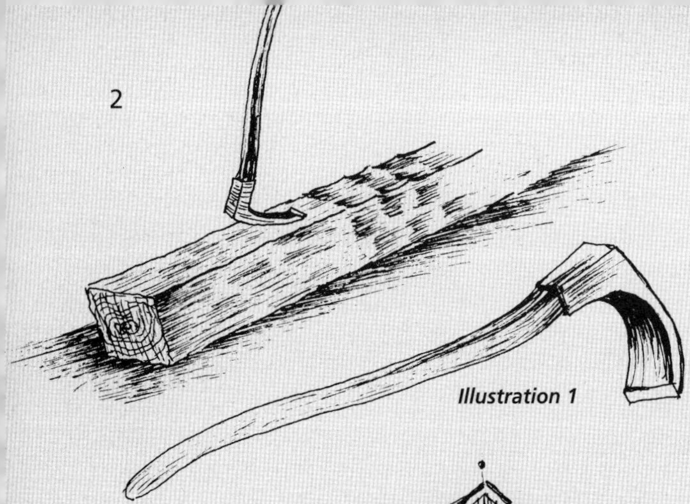

Illustration 1

Illustration 2

Illustration 3

The Language of Barn Building

ADZE: The adze was used for hewing beams, usually the smaller timbers, because it was a little more dangerous to use than an axe. With a handle almost like a pick, but with a graceful curve to it, the adze was used to chop away the wood from a round log. Straddling the log, with the adze sweeping past between his legs, many a hewer bore the scars of an adze on his shins. Illustration 1

ASHLAR PIECES: short upright timbers fixed to the wall plate and supporting the rafters.

BANK WALL: a barn foundation that is set into the side of a slope. The wagon doors are set on the bank side, with the bank acting as a ramp. Illustration 2

BAY: the space in the barn between two bents. Illustration 3

BEETLE: a large wooden mallet, its head made from ironwood. The handle is long and there are two iron bands around the ends of the head. Made for driving in the oak pegs or large beams into connections. A maul. Illustration 4

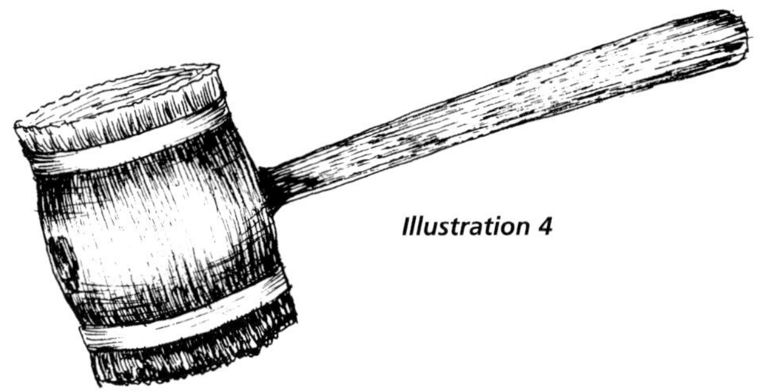

Illustration 4

BENT: one of a number of cross-sectional timber frames in a barn designed to support both vertical and horizontal loads. Illustration 5

BORING MACHINE: a hand-operated device with gears that drive auger bits for boring large holes in timbers, usually for mortise joints. Illustration 6

BRACES: short pieces of small dimensional timbers connecting an upper and lower timber at an angle of 45 degrees. Illustration 7

BROADAXE: The broadaxe preceded the sawmill. It originally came from Europe having been invented in the Middle Ages as a carpenter's tool. Short-handled and weighing about 10 pounds [4.5 kg], it was used for making round logs square. The process of "squaring up" a timber was called *hewing*. Deep notches were made, six to eight inches [20 cm] apart, on a felled tree, and then the broadaxe was used to hack out the wood between. This made the surface flat. Hewing continued until all four sides were finished, resulting in a square beam. Illustration 8

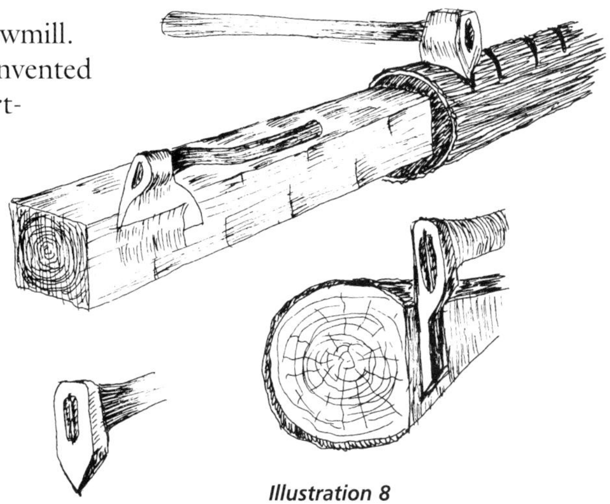

Illustration 8

COLLAR TIE: horizontal timber connecting two rafters to reduce sagging or spreading of the rafters.

CONNECTING GIRTS: horizontal timbers connecting the bents of the barn together.

END SILLS: horizontal timbers seated on top of the end of the foundation wall, supporting the floor ends and the end bents.

FOREBAY: (Laube) area under barn frame which extends beyond foundation wall as an overhang. Illustration 9

GIN POLE: a tall pole, often the trunk of a young tree, used as an aid in raising bents. With the gin pole set between the bent and a team of horses, the rope went from the bent over the gin-pole pulley and to the horses, which gave the proper angle of pull to raise the bent.

Illustration 5

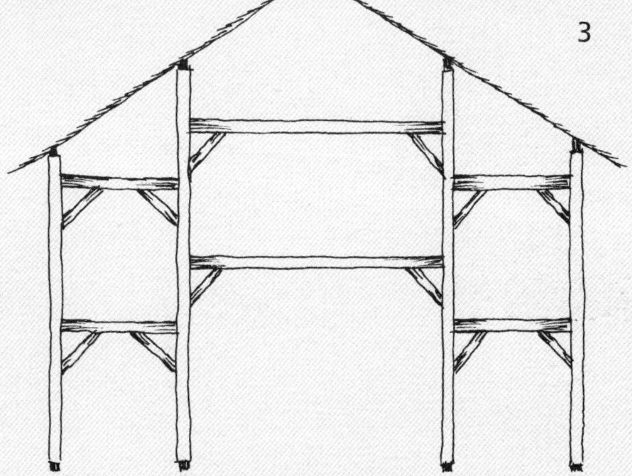

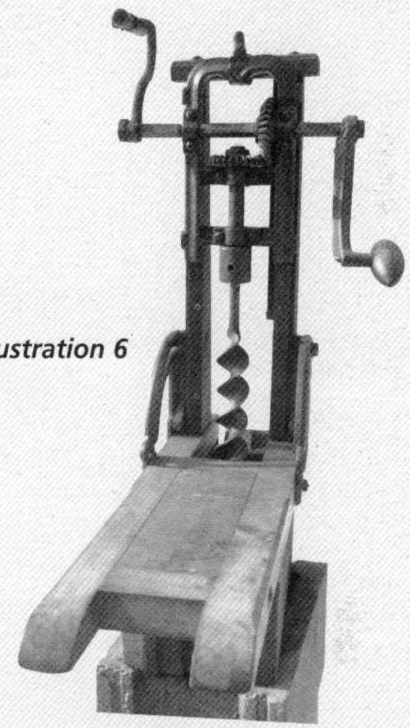

Illustration 6

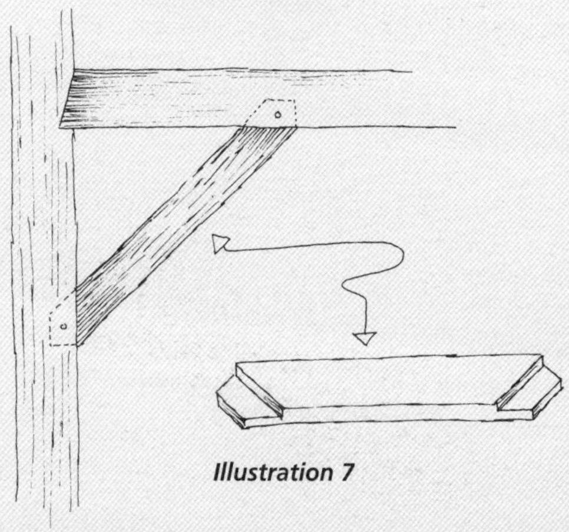

Illustration 7

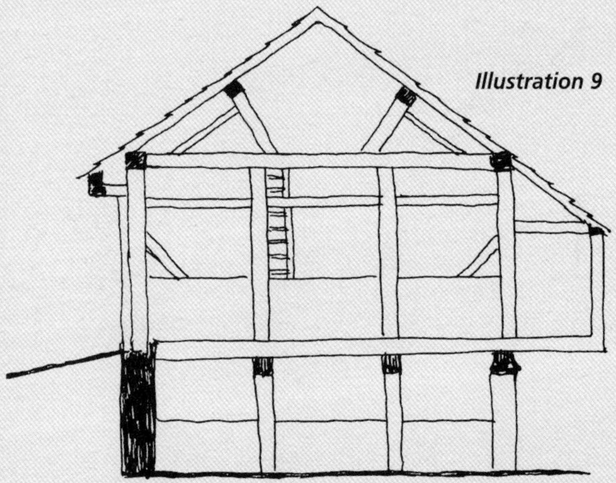

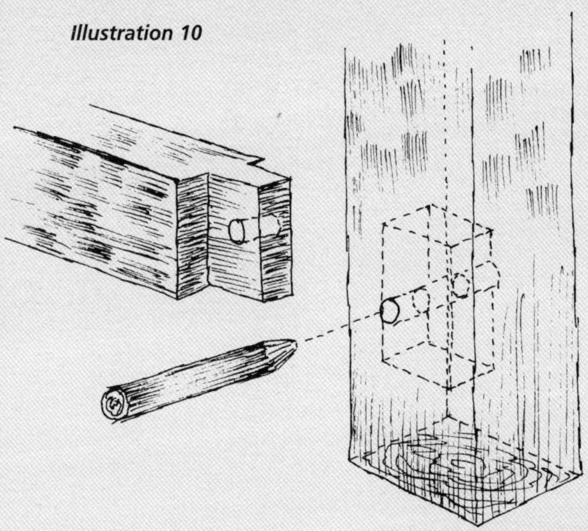

GRANARY: a room inside the barn's main floor containing bins for storing threshed or husked grain.

GRIST: ground grain used for feeding livestock.

KING POST: a timber roof truss which consists of a central vertical post from the bent plate to where the rafters meet at the ridge.

MORTISE: a rectangular slot drilled and then chiselled into one timber to receive the tenon from another. Illustration 10

MOW: the main storage area of a barn where hay, straw and sheaves of grain were stored.

MUD SILL: horizontal timbers resting on the ground, or on short pilings, rocks or posts.

PEG OR PIN: a wooden dowel from one to two inches [2.5 cm to 5 cm] in diameter shaped usually from oak, to act as a nail in pinning the different parts of the barn frame together. Illustration 10

PIKE POLE: a long pole with a metal spike attached to one end, used as a tool for pushing up the bents. In a barn raising sometimes hundreds of pike poles, of different lengths, supported each bent while it was lifted into place. Illustration 11

PLATES: major horizontal timbers that support the bottom of the rafters.

POST AND BEAM: another term for timber frame

PURLINS: within the roof structure, horizontal timbers that support rafters.

QUEEN POST: two vertical timbers of a roof structure placed on a bent plate or girt to support the rafters or collar tie.

RAFTER: timbers that run between the top or ridge of the roof and the bent plates or girts supporting the roof sheathing. In early times, round poles were used, or they were flattened on one side; later rafters were milled or sawn.

RIDGE: the very top of the roof.

SILL: horizontal timber resting on or fastened to the top of the foundation. Illustration 12

SLEEPER: a horizontal timber flattened on one or two sides, (often with bark left on) acting as a joist to carry the barn floor.

STABLE: lower portion of barn used to house livestock. Illustration 9

STALL: a compartment in a stable used to house animals in closer confinement.

SWING BEAM: a large uninterrupted beam spanning the threshing floor with no supporting posts between the walls. The purpose of this beam was to carry the heavy overhead mows and leave an unobstructed floor. Most often the largest timber in the barn.

TENON: the tongue chiselled at the end of a timber so that it will fit into the rectangular hole of the mortise. Illustration 10

THRESHING FLOOR: the area on the main floor of the barn used for unloading crops used for feed or for hand-or machine-threshing. The threshing floor was accessible from the main barn doors.

TIMBER FRAME: a frame made from large timbers which are joined and pinned together supporting smaller timbers on which the walls, roof and floor are affixed. Same as post and beam.

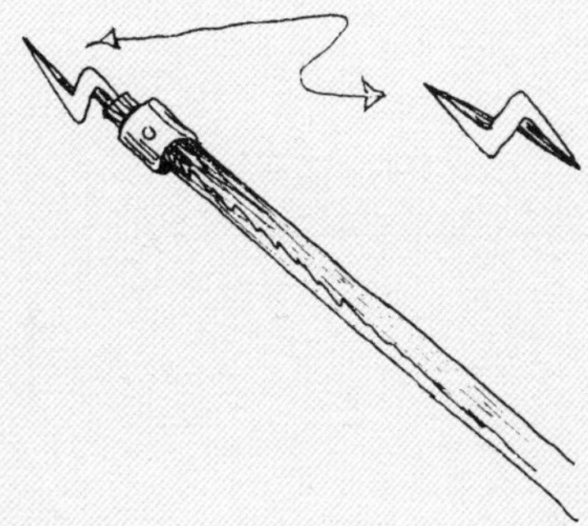

Illustration 11

Illustration 12

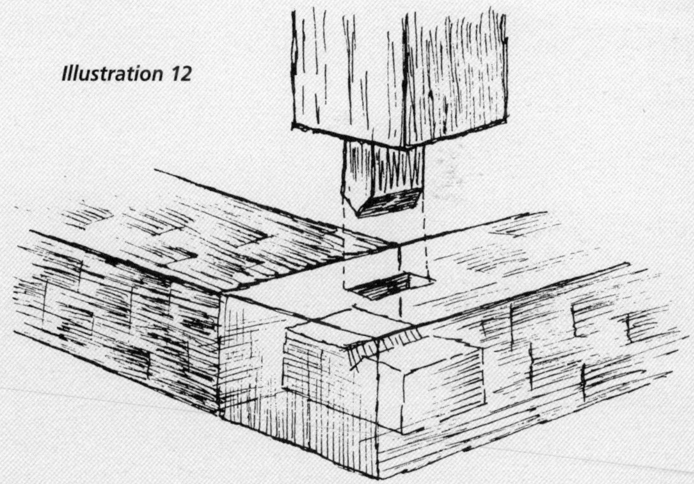

Barn Raisings

Barn Raising

THE CLARKSBURG REFLECTOR, May 28, 1903.

There was erected last Thursday and Friday on the farm of Mr. Chas. Parkinson, 8th Line, perhaps the largest frame work of any barn in Collingwood township. Mr. Parkinson, in remodeling his old barn, places on an enlarged foundation an immense structure 50 x 73 feet, with centre posts 26 feet and outside posts 16 feet long, with a 40 x 40 feet straw shed to the east side. The main building has a stone wall 9 feet high underneath four sides, while there is a substantial base wall underneath the straw shed.

Barn raising near Southampton, date unknown. Note the owner and his wife holding hands. (courtesy of Glen Stephens)

There is an old joke in the farming community—"The house may not be much," say the old-timers, "but the barns sure are nice!"

Barns held a special place in the pioneer farming community. The building of the structures took centre stage for the early farmer. Ontario barns differed significantly from the barns of Europe and Great Britain. Our tall barns were made of massive timbers and with steeply pitched roofs to endure the cold, snowy winters and sometimes vicious summer storms. Our barns had to store huge quantities of loose hay, straw, and grain, and keep livestock comfortable and warm in the winter. Building these mammoth structures was a huge task yet also a work of art.

Building a barn began with choosing a site, preferably a nice dry rise in the ground. Part of this hill was then scooped out to a depth of eight to ten feet [2.4 to 3 metres] so that a ramp could give access to the second storey and livestock could enter the stabling from below. Building into a side of a hill also helped to keep out winter's cold snows and summer's heat.

Stones had to be gathered for a foundation and there were plenty of those in the Queen's Bush. Limestone was abundant, as was sandstone, but most prized were the sparkling red, green, and grey granite stones. Those were usually reserved for the front of the foundation where they could be admired or dressed by skilled masons from Scotland and Germany and placed at the corners for strength.

To bind the foundation together, quick lime was bought from those with homemade lime kilns. This was limestone that had been burned in a pit, the heat cracking the stones and burning off impurities leaving almost pure calcium. When soaked in water it turned into a smooth white paste. Mixed with fine sand, the lime became a good binding agent. It dried as hard as the original stone from which it came. Evidence of lime kilns can still be found at many sites in the Queen's Bush, such as Lot 5, Concession 4, Greenock Township.

After completing the foundation for the barn, the farmer had to cut the massive beams for its frame. There was an abundance of trees on a pioneer farm. Many of these were of great size, sixty to one hundred feet tall [18 to 30 metres] and two to three feet [60 to 90 cm] in diameter. Species such as elm, maple, beech, and white ash were the best, but pine, cedar, hemlock, and tamarack often found a place in the barn frame. Usually the farmer cut the trees in the winter when there was a minimum of sap. Then a barn framer, such as Joseph Wrightson of Glammis, sent in his crew to hew the trees into workable timbers.

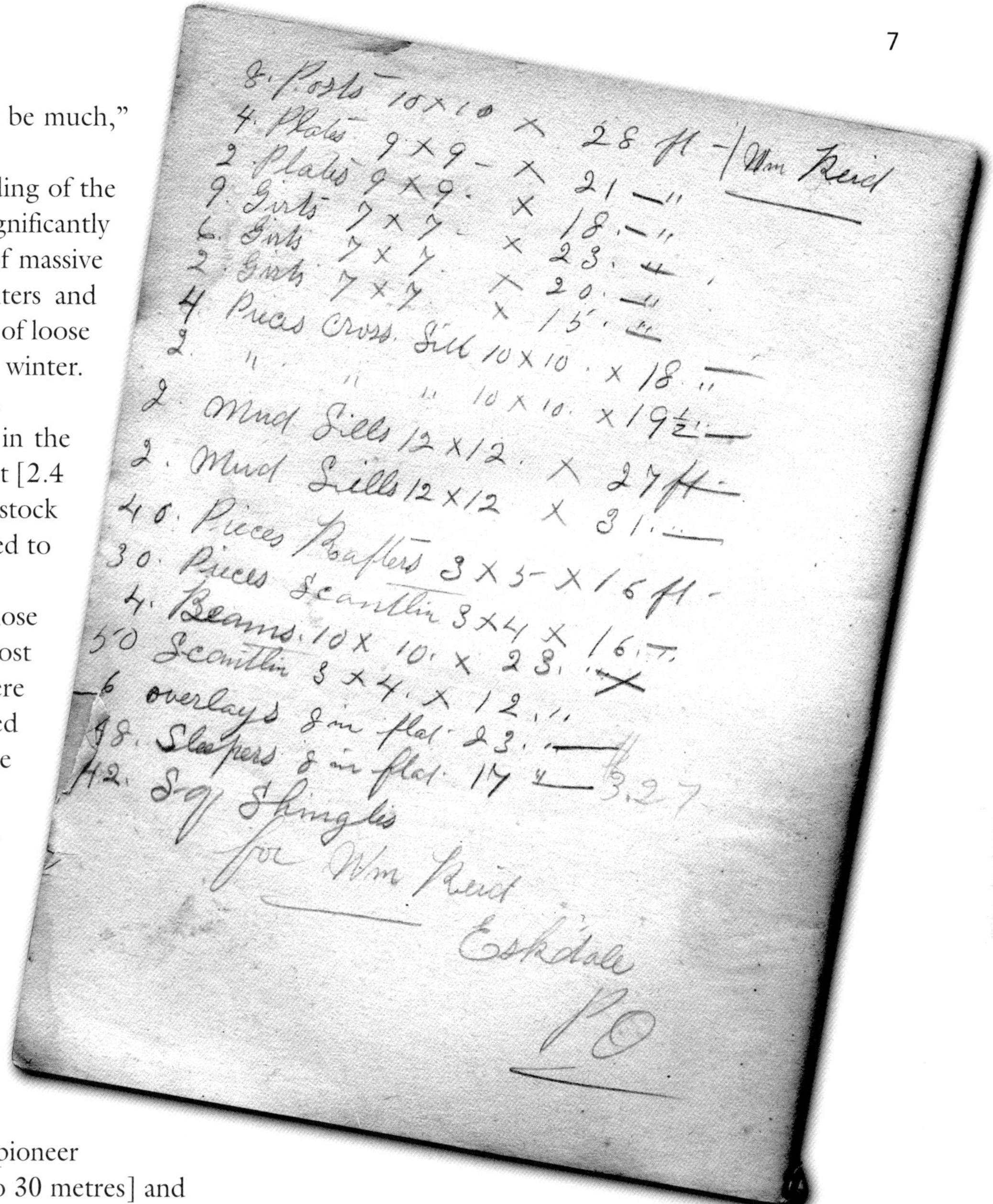
8. Posts 10 x 10 x 28 ft – Wm Reid
4. Plates 9 x 9 –
2 Plates 9 x 9 x 21 –
9. Girts 7 x 7 x 18 –
6. Girts 7 x 7 x 23
2 Girts 7 x 7 x 20 –
4 Pieces Cross Sill 10 x 10 x 15
2 " " " 10 x 10 x 18
2 " " " 10 x 10 x 19½
2 Mud Sills 12 x 12 x 27 ft
2. Mud Sills 12 x 12 x 31 –
40. Pieces Rafters 3 x 5 x 16 ft
30. Pieces Scantlin 3 x 4 x 16
4. Beams 10 x 10 x 23
50 Scantlin 3 x 4 x 12
6 overlays 3 in flat 23 –
48. Sleepers 3 in flat 17 –
42. Sq Shingles
for Wm Reid
Eskdale
PO

A page from the working notebook of Joseph Wrightson from Bruce Township, a master barn framer. (courtesy of Bruce County Museum & Archives)

Two men squaring a timber for a barn on the farm of Milton Moore near Ayton. (courtesy of County of Grey-Owen Sound Museum)

With a cord covered in red lead powder, the hewer would strike a line from one end of the log to the other, to give him a straight line to work from. This is where a strong back came into play. A "hacker" with a sharp axe stood on the log and struck the side of the bark every foot or so, down to the red line. The hewer would then stand beside the log and trim the slabs of wood down to the red line, making one side flat. Hewers needed considerable skill to keep the 10-pound [4.5 kg], one-foot [30 cm] blade of the broadaxe straight, despite varying grain and small knots in the wood. The log was then turned with its flat side down, and the whole process repeated. Looking at barn frames today, you can still see the marks of the broadaxe—some timbers are smooth, with hardly any marks, hewn by an expert, while some have deep cuts still showing, perhaps the marks of an apprentice. A variation of this method was the use of an adze instead of a broadaxe for the final finish.

The timbers were then gathered in front of the barn foundation and the framer began the complex work of marking and notching each beam. The crew arranged the timbers in the order in which they would be used. Using a framer's square, the master framer marked out each timber. Mortises or boxes were marked and tenons or tongues to fit them. There were gains or slots for the ends of the rafters to fit into, and different mortises for the braces used at almost every intersection of vertical post and horizontal beam, to strengthen the barn against winds and snow load. Allowances had to be made for timbers shrinking or twisting, so that the whole process of marking out and notching a frame had to be meticulously done, even to the point of using a metal scriber to mark the lines because the width of lines from a dull pencil could cause a misfit.

It was important for the notcher to closely follow the lines left by the master framer. The hardest part was boring the holes for the mortises. At first this was accomplished by using a hand auger with a bit two inches [5 cm] in diameter and about two feet [60 cm] long. A long stick or metal pipe was placed through an eye at the top to form a handle and then the auger was turned by hand to bore the hole. Mortises could be as deep as one-half the depth of the beam. Imagine twisting that auger into rock elm or hard maple ten to twelve hours per day!

A boring machine.

A later variation was a boring machine with an auger built into a wooden frame. The boring machine sat on the beam and as the workman straddled the beam, he turned the handles, like bicycle pedals, with his arms. The auger was forced down by the screw at its tip. Both these methods used pure human muscle power, repeated over many long days. The wood left by the auger was removed with a chisel and mallet.

The tenons were made either by using the axe to chop down to the required line, then taking off the wood with a chisel, or by using an adze to clean away the wood around the tenon.

When all the mortises had been bored and chiselled, the tenons shaped, the braces cut at the right 45-degree angle and pin holes bored through the junction of each joint, it was time to assemble the pieces into bents. Bents were sections of the barn frame fitted together on the floor of the barn, prior to the raising. In fact, it was these great sections—bents—that the raisings were all about. Some smaller barns had only four bents; some large ones had as many as eight. The bents were joined to each other by connecting girts. When these were all raised together they became the massive barn frame.

A little time for fun at a barn raising in Wellington County. (courtesy of Mary Tripp MacCarl)

Assembling of the structure began with the sill plates, which were man-handled by the crew to the top of the foundation. These beams held not only the floor, but the posts for the main frame. Next, the interlocking joist timbers, called sleepers, for the main floor, were laid on top. Many of these were just flattened while the timbers were freshly cut, on one or two sides, with either a broadaxe or adze. It was much easier to square up or flatten logs that were newly cut and not hardened under a summer sun.

The threshing floor especially needed many sturdy sleepers, more closely spaced together, as this was where the wagons rolled in, full of hay or grain, pulled by a pair of large draft horses. The floorboards, made of whatever wood was near the farm, were often two or three inches thick [5 cm to 7.6 cm] and as long as possible. They would have been cut at a nearby sawmill.

Once the farmer or a hired framer had the deck of the barn ready, the actual assembling or barn raising would be called. Every barn made in the Queen's bush was distinct. Size, width, length, height, and type of wood all differed, yet the overriding similarities in the minds of the pioneers of the nineteenth and early twentieth centuries were solidity, durability, permanence. They built for perpetuity—a word we don't hear often in today's fast world.

A barn raising was something special in the community. Men, women, and children would arrive for the big day,

Barn being finished, before 1900. (courtesy of Bruce County Museum & Archives)

Barn with block foundation being finished in the 1920s. (courtesy of Bruce County Museum & Archives)

which was also a major social event. The framer would be there with his crew and a great assortment of tools—chisels, cant-hooks, adzes, axes, mallets, pry-bars, augers, crowbars, and pike poles of various lengths.

All the bents would already have been assembled and laid out on the floor of the barn. Usually, two captains were chosen, men with experience at raising barns; each in turn chose his team for the great effort. After the two teams took their positions, the raising would begin, soon becoming a friendly race to see which side would finish first at their assigned part of the barn.

Through sheer manpower and a series of lifts guided by yells, the heavy frames would be lifted off the floor or ground. With the sharp ends of the pike poles sticking into the frames, a final mighty heave raised the bent into place. It was propped there until the connecting girts were put in and then the men swarmed over the frame, inserting the braces and pounding in the wooden pins that held everything in place.

Proud People, The Lindsay Township History Book, includes a recollection of a barn raising by William Gilroy, the young Methodist minister at Dyer's Bay in 1897-8:

> One incident that stands in my mind was the raising of the largest barn ever raised in the area. Captains and sides were chosen and a wild race began (to see who could put up their side of the roof rafters first). Indescribable pandemonium broke out with ordinary quiet church-going men swearing and cursing in a reversion to wild lumbering days. The side I was on was well ahead when a purlin plate came apart. I was among those holding the heavy weight on a pike pole with the threat of the structure falling. The uproar was finally stilled and a man went up to drive in the loose peg. But by that time the other side had won. When we went later to wash up at the pump I said 'I'll tell you boys what was the trouble. You had the minister on your side, and you had one man less to swear.'

Purlins, beams equalling the length of the roof, were difficult to manhandle up to the top of the frame. Held in place by long braces, or by the posts themselves, the purlins' main job was to support the rafters of the roof. Once the rafters were installed the barn raising was considered complete. The farmer, with the help of a smaller group of neighbours, finished the barn siding, roof sheathing and finally cedar shingle roofing on top. On a rainy day the granary would be built inside the barn. Later, the farmer completed the gangway up to the main barn doors.

After the raising was done—usually at sunset—the men would sit down to a large meal, prepared by the woman of the house and several neighbouring ladies. The food would have to be both generous and delicious, for hundreds of dollars worth of labour had just been donated to the farmer and his wife, and a good helping of generosity would be offered in return. From *A History of Normanby Township,* here is a list of food for a barn raising involving 175 men: 115 lemon pies, 500 doughnuts, 15 large cakes, 3 gallons apple sauce, 3 gallons cornstarch pudding, 16 chickens, 3 hams, 50 pounds roast beef, 300 light rolls, beet pickles, pickled eggs, 6 pounds dried prunes, 1 large crock stewed raisins, 5 gallon stone jar white potatoes, 5 gallon stone jar sweet potatoes, and cucumber pickles.

You can hear the excitement in the voices of those who can still recall barn raisings.

Franklin Jackson was born and raised on a farm in Glenelg Township. This is one of his experiences during the late 1930s at a barn raising for someone whose barn had burned down. "We'd be there after chore time, by 8 o'clock. The foreman would have everything ready—the big sledge hammers, a large assortment of pike poles, all that. The whole neighbourhood was there, about 150 people, and there was certainly a buzz of excitement in the air. At the call of the foremen we'd all lift the bent by hand and some would place short pike poles under, about four feet high, to prop the thing up. And then some of us would lift the structure to the top of our heads and the others would then insert the little bit longer pike poles and push it up another four or so feet. Then, as they were holding it others would start pushing with even longer pike poles and so on until it was up," explained Jackson in his kitchen one fall morning.

"Once the frame started coming together it would be like a swarm of ants. People all over the frame, pounding, pushing, lifting. It was neighbours helping neighbours."

It was a tremendous task to prepare food and serve so many hungry and thirsty men but the raising was also a gala time. Barn raisings provided a chance to meet old friends and exchange news and a time for young men and women to look for their lifetime partners. Often, a barn dance followed the meal, with a fiddler and a caller and sometimes plenty of homemade beer or whiskey. Couples would swirl around the new floor, looking up through the barn frame to the starry sky.

The barn raising was one of the cornerstones of community sharing and living. Everyone chipped in and you knew, if your own chips were down one day, there was a community out there to help you.

The great era of building timber-frame barns in the Queen's Bush lasted from 1860 to 1940. After 1940 there were hardly any large trees left uncut, and other forms of construction, such as the steel framed "kit barns" and the pole and stud frame barn, took over.

The Barns

1. Many Barns in the Making

More than 120 years ago, farmer John Mackenzie had the foresight to plant trees around the perimeter of his farm. Today, the majestic maples hug the road edges of the corner farm situated in the heart of Sullivan Township, about two miles west of Chatsworth. This man also took on the daunting tasks of building four barns, taking down a couple, moving and reshaping them to his farming needs.

John also had something most farmers in those pioneer days didn't have—enough money to do it right. With cash in hand, he set about building a model farm.

By 1880 John Mackenzie had built many of the outbuildings and barns he would need. Twenty years later, the barn in the foreground and the barn on the left in photo 1 had already been taken down or moved. Neither one had foundations and their use could have been more as driving or straw sheds. Remaining today are the two barns we see in photo 2; the barn in the foreground had a hip roof added in the 1940s and the other one was raised and a stone foundation for stabling added between 1870 and 1900.

1-1 *Four barns, a windmill and a 2,600 lb [1,180 kg] shorthorn bull that owner John Mackenzie imported from Scotland in the 1870s to start his breeding program. The dog is watching the sheep, a breed so woolly they had trouble seeing because of the amount of fleece over their eyes. (courtesy County of Grey-Owen Sound Museum)*

Around 1880, Mackenzie imported from Scotland, the country of his birth, a shorthorn bull for breeding purposes and three Clydesdale mares. The bull, bred with local stock, produced dual-purpose heifers, good as beef cattle and also prolific milkers, while the horses took over the work from teams of oxen.

Brian Gamble, the present co-owner of the farm, which was originally purchased by his father Mac in 1964, found bits of history left behind by John Mackenzie. There is a manual for an A.A. Cochrane reaper-binder from 1875, a bill from 1881 for fire insurance for the farm totalling $2.81, and a notice showing taxes for the year 1880 at $54.

Mackenzie seemed to keep the farm on the very edge of progress. "He really started doing things," said Brian, of the stories passed down about Mackenzie. He installed a Cooper windmill to be able to grind grain inside his barn. The early water pumping windmill was replaced by a much higher one in the early 1900s and, about the same time, he installed a Delco system, an early electrical generator of direct current (DC), as opposed to our present system of alternate current (AC).

1-2 *It's threshing time at the Mackenzie barn in 1900. A steam-powered engine is running a thresher inside the barn. The Cooper windmill is connected to a grain grinder in the mow of the barn. (courtesy of County of Grey-Owen Sound Museum)*

In 1908 Mackenzie hired a crew to build a silo. It was 12 feet [3.6 m] in diameter, over 28 feet [8.5 m] high and had six-inch- [15 cm] thick walls of cement. Brian Gamble knows how solidly this silo was built because when he eventually wanted to take it down, a large bulldozer worked many hours to topple it. Even now, parts of the bottom of the silo lie embedded in the ground, a testament to the "forever" thinking of the builders.

In 1900, Mackenzie and his neighbours started one of the earliest beef rings in the area. The building still stands, a small shack attached to the driving shed. Inside is a big wooden wheel, about 8 feet [2.5 m] in diameter, used to slowly raise the beast to allow easy butchering after it was slaughtered on the cement floor below. Brian Gamble remembers seeing pieces of paper in the shack, held to the wall by clothespins, with the names of neighbours involved in the ring.

The beef ring continued after John sold the farm to George Sutherland in 1924 because it was a sensible community solution when freezing meat in the summer was not yet possible. It also went hand in hand with community sharing. During the difficult thirties, beef rings were the only source of meat. Local farmers got together and formed groups of around sixteen shares. Depending on their needs, an individual could own a full share, a half-share or a quarter-share. Members of the ring would be elected as officers—a chairman, secretary, two inspectors and two auditors. The inspectors' duty was to examine the animal two weeks before slaughter. One beast was contributed per share on a rotating basis.

The cuts of meat were allotted in order, so that each family or shareholder would receive different cuts of meat every week. Each share provided two steaks, two roasts, and two meats for boiling. A quarter-shareholder would

1-3 *In 1908 John Mackenzie (standing in the centre holding a horse) hired a crew to build a 28-foot by 12-foot [8.5 m x 3.6 m] diameter cement silo which was located where the two barns met. A cement mixer sits on the wagon with a motor off to the side. Since John did not marry until in his sixties, the young family sitting around the wagon on the right are probably neighbours. This is a vivid snapshot of what life was like 100 years ago. The barn behind the silo is the hip-roofed barn we see in photos 4 and 6. (courtesy of County of Grey-Owen Sound Museum)*

1-4 *Another comparison 100 years later. The barn has a hip roof added and is connected to a second barn on the back and a low building to the side. Gone are the Cooper windmill and the cupola on top.*

receive a steak and a boil one week and a roast the next. Each member's meat was weighed and recorded and then placed in a basket for pickup. At the end of the year the auditors would check the weight of each animal brought in against the weight of the meat received. A difference would require the shareholder to pay for the extra he received; if he was on the short end, the beef ring would owe him some meat.

Mackenzie must have been an exception when he planted trees around his farm. Pioneer pictures of this area show a surprisingly barren-looking landscape. It is said that by the early 1900s the forests were cleared from Mount Forest to Tobermory. The pioneer farmer needed fields to grow crops, and trees were considered an obstacle. In Sullivan, a person could receive a free 50 acres [20 hectares] of land if he cleared and raised crops on one-third of it. All this had to be done within four years. There is an account by Mrs. Gottlieb Guse of coming to Sullivan with her husband in the 1870s and thinking she was approaching "a big, brilliantly lit city." It turned out to be burning stumps glowing everywhere over cleared fields.

Mackenzie was a builder. His barns were solidly constructed of hardwood beams, and the two

1-5 *The 8-foot [2.4 m] diameter wooden pulley and axle used to pull up slaughtered cattle for butchering. It was utilized well in the beef ring which operated on the farm from 1900 until the 1940s.*

1-6 *The other end of the hip-roofed barn, which was remodelled in the 1940s to accommodate more loose hay and straw, just before square bales became popular in the '50s.*

barns we see today are large compared to others built in the same era. The first barn was a standard 40 feet by 60 feet [12 m x 18 m] with a queen post roof system and 60-foot [18 m] top plates made from elm spanning the length of the barn. In the 1940s this roof was altered to a hip roof to accommodate more loose hay and straw. It's easy to see the timbers used in this expansion were of a later date, because of the saw blade marks on them, compared to broadaxe marks on the older ones. Most of the timbers in this barn were white ash and the posts were a substantial 12 inches by 12 inches by 22 feet [30 cm x 30 cm x 6.7 m].

The second barn we see in photos 7 and 8 may have been the first one built, but was greatly expanded between 1900 and 1907. It was split in the middle and the roof raised making it 70 feet

1-8 *This barn extends from the hip-roof barn in photo 4. It was split in the middle at the beginning of the last century. Another 30 feet [9 m] were added to the width to make the barn a full 70 feet wide [21 m]. The sixty-foot [18 m] long structure has a very different bent configuration which does not meet in the middle, only on the ends. (See illustration below)*

1-7 *Much has changed in the 120 years since the first photograph was taken. Still standing is the barn on the left, with a hip roof added.*

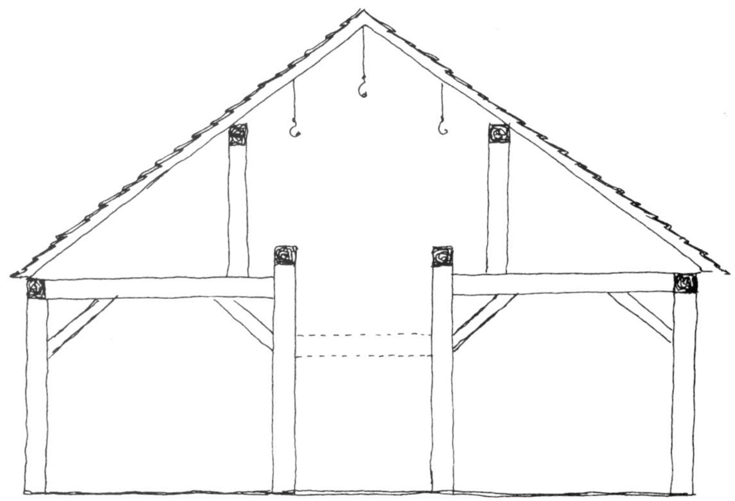

wide by 60 feet long [21 m x 18 m]. An ingenious 5-bent timber framing system was used to build the expansion. There were no crossbeams in this configuration, except on the ends, so that the use of the hayfork was greatly enhanced. As a matter of fact, because the barn is so wide, three hayforks were needed in this great structure that is over 45 feet [14 m] high. One is in the center at the ridge and two are on either side about 10 feet [3 m] away, running the 60-foot [18 m] length of the barn without any obstructions.

The great amounts of hay that were stored in those barns came from the fertile, slightly rolling land surrounding them. It still looks the prosperous farm, as John Mackenzie envisioned it; the great maples he planted add a stately charm he would have been proud of.

1-9 *The maples that John Mackenzie planted around the perimeter of his 100-acre [40 hectare] farm in the early 1900s give a certain grace to the holdings.*

2. The Moveable Barns at the Fligg Farm

No less than three barns were moved to this homestead during the 130 years the farm stayed in the Fligg family. Two are attached to the home barn and one stands by itself, perched on one of the many hills in the area.

In 1860 William Fligg Sr. came to this St. Vincent Township farm, located about 7 kilometres west of Meaford, and took out 50 acres [20 hectares] from the Crown. He ran a blacksmith shop a mile from his farm in the village of Woodford. As soon as he settled down he travelled to Newcastle, just outside of Hamilton, married, and brought his bride back to Grey County. Family history recalls that on the return journey, the newlywed couple lost all their possessions and arrived at the farm with just the clothes on their backs. Travel was very difficult then and there were many opportunities to lose a wagon in one of the swampy roads leading to Collingwood. From there only one road linked them to the St. Vincent farm.

In fact, Woodford was the end of the line for horse travel because of the miles of rough outcrops of bedrock that restricted travel west of the village to a footpath. This is why in later years the train tracks ended at Meaford, with no connection to Owen Sound. The main road from Meaford to Owen Sound followed the peninsula through the town of Cape Rich, which in 1940 was part of the 17,000 acres [6,880 hectares] taken over by the federal government for the Meaford Tank Range.

William Fligg Sr. was an enterprising man. He bought three adjacent 50-acre [20 hectare] farms, with barns and log houses, to expand his operation. The property is upland and the proximity to Georgian Bay has a warming effect on the area although parts of the land are swampy with bursts of bedrock here and there.

2-1 *This is the barn that was moved almost 100 years ago. It stands by itself because the owner, William Fligg, was afraid that one barn fire could set all his other barns ablaze.*

Eventually the distance between the different barns became too difficult to manage. The barns were mostly on mudsills with no stabling below, and were used mainly for storing loose hay, straw, and threshed grain. Two of the barns were moved and connected to the

2-2

2-3

2-4

home barn which had been built in 1889. The third barn was moved at the beginning of the twentieth century by William Fligg, Jr., who took over the farm from his father. This barn was a kilometre away, at the next farm east of the Fliggs' place.

The 36-foot by 42-foot [11 m x 13 m] barn was pulled, nice and slowly, by one horse attached to a windlass, a simple winch with a crank, over the mostly flat, sometimes wet fields, and then up a fairly sharp hill to the waiting foundation. Large trimmed hardwoods were used as rollers under the barn and, as the barn was moved, each roller popped out at the end and was taken around to the front again by another horse, to be inserted under the barn's mudsills. The queen post framed barn, with its four bents and a granary was winched on top of a hand-poured cement foundation.

It's hard to imagine buildings of that size being moved across rough fields, with just horsepower, patience, and ingenuity. Picture these massive structures, swaying over the rough bumps, held together with only wooden pins and cedar braces.

The small barn's three-post bents were of a classic, simple design, used by many farmers and framers throughout the Queen's Bush. Since there were no splices in any of the bent girts, connectors or purlin plates, the beams of white ash and maple were either

2-2 *The barn was moved about 1 kilometre across this bit of rough swampy ground. Its original location was near the house on the right side of the photograph.*

2-3 *It must have been quite a sight to see a barn moving across a field. Many neighbours were on hand to watch or help. Note the windlass is being turned by the horse as someone leads it. The barn boards look fresh and the roof straight. (courtesy of County of Grey-Owen Sound Museum)*

2-4 *The barn was pulled up this hill to the foundation. It had to be pulled past the foundation, jacked up and then rolled back onto the cement support.*

36 feet or 42 feet [11 m or 13 m] in length. Twelve, one-piece posts were 20 feet [6 m] in height to the bottom of the rafters. Timbers were simply cut to length and the ends notched, making the framer's job simpler.

The barn was settled onto a foundation about 100 metres from the main barn and a little farther from the house. William Jr. was taking no chances. Barn fires were a farmer's nightmare so he placed this barn away from the others, thinking if the main barns burned down he would at least have this one left. It made for longer walks during chore time, but it must have given William Jr. peace of mind.

Like many others in the Queen's Bush, what made this farm successful was family and neighbours. From the original pioneer, William Sr., to his son William Jr. and his son William Harvey to his son, Wayne, who still owns 25 acres [10 hectares] of the original property, the various uncles, aunts, sons and daughters, grandfathers and neighbours, every year got together and pitched in to do each other's crops.

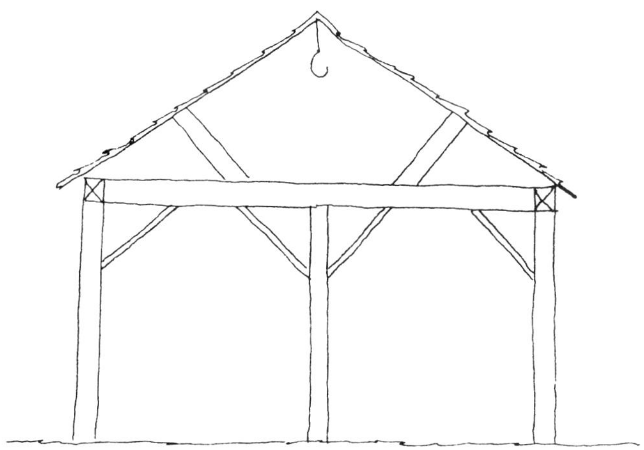

2-5

2-6

2-5 *The barn boards probably have not changed since the barn was moved as they show the same lines as in the original photo.*

2-6 *Inside we see the simple three-posts-per-bent timber frame and queen post braces for roof support.*

2-7 *The main doors of the barn and ramp lead up to the threshing floor.*

"I remember the threshing crews. I was only five years [old] at the time, and I would drive the horses pulling the wagon piled high with stooked grain. All the family was there—my dad, uncle, his sons, and neighbours. We would take turns doing each other's farms," said Wayne, who sold most of the farm, including the barns, in 1990.

"Now farms are set up for a one-man operation and you still can't make any money," Wayne added. With his wife, Penny, and daughter, Robin, he now cash crops on mostly rented land in the area, works a full-time job in Collingwood, and served on St. Vincent Township council for many years.

The Fligg farm grew a variety of crops between the 1920s and the '50s: wheat, fall barley, mixed grain, alfalfa, sweet clover, turnips, potatoes and most of the vegetables the family would need over the year.

Wayne remembers grain was threshed during the beginning of winter in each barn. The stooks were brought in by wagon and then the wagon deck was lifted up with a large wooden pulley in the barn and the stooks forked on top of the hay. After all the fall work was done the Fliggs' family-owned thresher would come around to each relative's barn and the work of threshing the grain would begin. Their first thresher was a Frano and the last a McCormick, which Wayne used until he sold the farm. "The stooks of grain would get beautifully dry by the winter and when threshed, it was the best grain," Wayne said.

Pigs and sheep were kept in the small barn and Wayne spent a lot of time with his father, William Harvey, showing their registered Yorkshire pigs at many of the fairs in the area. The prize-winning pigs were shipped as pure stock across Canada and the USA.

Wayne, in his late forties today, was fond of horses and used them around the farm until 1980. His father owned Big John, a Belgian workhorse that was reputed to be the biggest horse in the world at the time, weighing in at 2,640 pounds [1200 kg]!

It was a different community in the early twentieth century. The land was supporting a much bigger rural population, with large families working every farm across the landscape. There were many more towns and villages. The little village of Woodford, which can be easily missed today while driving through in a car, had at one time a hotel, a mill, two general stores, a post office, a brickyard, a public skating rink and stagecoach stop, where the horses were changed.

This way of life may have changed a great deal, but evidence of its existence can still be found when we look at the many pioneer barns dotting the countryside.

2-8 *Stooked grain ready for threshing, much like in the days of threshing bees at the Fligg farm.*

3-1 *A close-up shows the pine siding looking worn and paper thin, but still defying the weather.*

3. The Pine-Framed Barn

This wonderful barn, built in the 1860s, has many exceptional features. It is very much in its original shape—even the horizontal siding has never been replaced. Particularly amazing and somewhat unusual to this area is the timber frame itself—a 40-foot by 70-foot [12 m x 21 m] frame all notched from one-piece timbers. Imagine the size of the trees used!

The barn was built in Bentinck Township northeast of Hanover by Lloyd Chittick's great-grandfather, who bought the 100 acres [40 hectares] from the Crown in 1854 for $160. To get full possession of the deed, the requirement of the Queen's Bush settlers was to clear at least two acres a year for up to three years. Imagine a forest of tall, predominately hardwood trees to fell with nothing but a crosscut saw and an axe, streams and swamps to cross and (if the farmer was fortunate) a team of oxen to help pull stumps and boulders out of the way. All this plus a house to build for the family as well as a barn to make a living from.

3-2 *The 40-foot by 70-foot [12 m x 21 m] Chittick barn with original siding from the 1860s.*

3-3 *The pine frame includes 14-inch by 14-inch [36 cm x 36 cm] posts 26 feet [8 m] in length and girts 40 feet [12 m] long. A hay rake and other machinery hangs from the ceiling.*

On this Chittick farm is a stream with softwoods growing around it. It was from here that the great trees were felled to construct the pine timber-frame barn still seen by those who drive up Grey County Road 3, two concessions north of Grey County Road 4. The posts, 26 feet [8 m] high are massive 14-inch by 14-inch [36 cm x 36 cm] timbers. There are two 70-foot [21 m] bent girts as well as two purlin plates of the same length. A swing beam measures 20 inches [51 cm] across its face and runs the width of the barn. Huge timbers, all of them from trees of great height cut down from virgin forests.

This barn was built as a storage space for hay, straw, and grain; there was no stable for livestock. The base of the structure is level with the ground and sits on massive mud sills, which are resting on a few boulders placed on three corners to level the barn. (Photo 5)

The original small frame house was already torn down to make room for an expansion of the county road when Lloyd bought the farm in 1952. An attached shed on the barn, built to house livestock, fell down 40 years ago.

Walking inside the barn feels like going back into history. Apart from some hay stored, there are many early implements such as turnip sowers, sleighs, cultivators and Lloyd's pride and joy, a 1932 Robert Bell thresher.

Lloyd bought the thresher in 1959, along with a 1941, 62 horsepower Minneapolis Model G tractor to run the thresher, for a total price of $3,955. The first year he threshed grain on 34 farms with the steel-wheeled Robert Bell. He continued to thresh for farmers until the early 1970s. He still uses the Robert Bell to do his own grain.

3-4 *The swing beam measures 20 inches by 14 inches by 40 feet [51 cm x 36 cm x 12 m]. It gave good access below the stored hay without any post restrictions. This was usually where the granary was located.*

3-5 *From the south it is easy to see the barn's mudsills, supported by a few boulders at the corners for the last 130 or so years.*

Lloyd's eyes sparkle when he talks about the thresher's long history. During the 1930s it routinely did 60 farms per season and the last custom job the Robert Bell did was in 1974, after 42 years of hard service. Lloyd learned well one lesson about threshing inside his barn—don't. The one and only time that he did, so many mice moved in afterward looking for the residue of grain, that they chewed through some of the Robert Bell's belts and pieces of leather. Now Lloyd waits for a few sunny days, hitches up the thresher to the still-running Minneapolis and tows the Robert Bell to his other barn, on the home farm, where he threshes the grain.

During the month of July, Lloyd brings out a Massy-Harris reaper-binder to cut and sheave the grain. After stooking the grain to dry in the field, Lloyd and his son Dave from Allan Park load the sheaves onto a wagon and bring them to the home barn, where the thresher is set up. One by one they pitch the sheaves into the thresher, as it separates the grain from the straw. The grain is blown through a pipe attached to the thresher into the granary and the golden coloured straw into the mow. (Photo 8)

3-6 *A 1939 Robert Bell thresher on steel being pulled by a 1941 Model "G" Minneapolis tractor. The "unit" has been a part of farming on the Chittick farm since 1959.*

"I never thought a combine was as good in this part of the country. It just doesn't work as well," Lloyd said. His wife, Evelyn, added they don't lose any of the straw, since it is brought to the barn and cut up inside the thresher, so that the short pieces soak up the animals' urine better.

Even though they think combines provide "some independence for farmers" this is offset by the cost of the machinery, which means farmers need off-farm work to provide the money for the big equipment. "You have to get steady work with the expense of the machine," said Evelyn.

Not too long ago, farming was a community affair. Threshing and haying, for example, were jobs tackled by the whole neighbourhood, including children. "Farmers used to work together on all the jobs and help each other if one was sick," said Lloyd, "but now, all you see are lonely-looking farmers on tractors."

Evelyn remembers cooking for threshing gangs of up to 14 hungry men. "It was a competition to see who put up the best meal. I remember one man telling me the meals were the best part of threshing," she laughed.

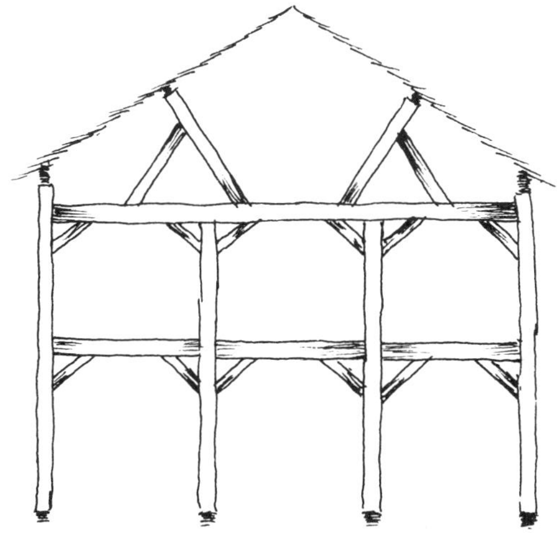

3-7 *Part of the workings of the Robert Bell.*

At one time the Chitticks kept pigs, laying hens, sheep, some dairy cows and beef cattle. All those were necessary to survive the depression of the "dirty thirties." They churned butter by hand and sold it for five cents per pound; they received three cents per pound for fat pigs and calves sold for $4.50 each. "Farming has changed so much. Young farmers want to start at the top but they get into so much debt that way," Lloyd emphasized.

The Chitticks still have about 40 head of beef cattle and a few laying hens. "He's never had to have a hired man. I've been out in the fields with him all this time," said Evelyn proudly.

The secret to keeping machinery so long is to have it well maintained, said Lloyd. He still uses the first tractor he ever bought—a 1948 Farmall Cub, which he purchased brand new.

The active participation that farmers traditionally took in the community led Lloyd inevitably into local politics. He served as Bentinck Township councillor on and off for 32 years, beginning in

3-8 *Everything is awhirl and filled with noise. The Minneapolis tractor is running hard, powering the belt attached to the thresher as Lloyd pitches in another sheaf.*

1948. The last 18 years he was reeve, ending his political career in 1988.

Lloyd tells stories from his job of 30 years as a clerk for an auctioneer. "Once a cheap farmer bachelor bought a hay block at the auction for only two dollars," Lloyd remembers. "When the thing didn't work I told him it wasn't any good because he didn't pay enough for it." For a while Lloyd had a 40-kilometre route picking up cream and eggs on a sleigh for a distributor. "We used to unload everything in the house so it wouldn't freeze," said Evelyn, "and then in the morning load it up again so Lloyd could take it down to the truck waiting on the highway."

"My great-grandparents brought up seven children on this farm over a hundred years ago, yet today I can't even make a living for two. What is going to happen to all the farmers?" asks Lloyd, who is in his early eighties. He farms 200 acres [81 hectares] one concession south of the pine barn, raising beef, still doing all the chores twice a day with his wife, Evelyn, and occasionally helped by his son, David.

It's the active part of farm work that keeps the Chitticks going when most people would be retired. But it's a love of the life on the farm that is behind the will to get up early on a winter's morning to do the chores.

Plans are to replace the almost 140-year-old clapboard siding on the pine-framed barn and keep it in good order, as there are some central posts that are sinking and pulling away from the mortise joints. How many more generations of farmers are willing to keep up the old barns is anyone's guess but, slowly and surely, the great timber barns are disappearing, and with them the attendant skills and a way of life just three short generations away.

3-9 *The main doors of the barn frame Lloyd, right, and his son, Dave. The siding is so thin in places that light is shining through.*

4. Log and Timber-Framed Barn

Stories of childhood country memories capture some of the strongest images of growing up on a farm. Pictures remind us of the happy times and tend to let work fade into the background. Sore muscles are forgotten and the photos show only smiling faces and family together.

So it is with this remarkable farm with its log barns mostly all collapsed. Although its roof has been shredded by the wind, one still stands, echoing the vibrant life of a family that once worked the soil and used the barns for work and play.

Log barns were the first large livestock structures that the pioneers of the Queen's Bush built. Clearing the land produced logs which were a natural building material to use. Many barns were hastily built, without a foundation and with greenhorn construction skills. As they deteriorated quickly, timber-frame barns replaced them and the log barns decayed into the ground.

4-1 *The barns as they looked in 1959. The long log barn in the foreground has all but collapsed except for the driving shed. The tall log barn is still standing but is no longer used. (courtesy of Marg Martin)*

So it is unusual that parts of this log barn, situated behind the former Glenelg Township office, about 12 kilometres northeast of Durham, have survived at all. The McClocklin family farmed the land from the early 1920s until the mid 1980s. "We were a family of hard workers and practical jokers," said Verna. She is one of six sisters and two brothers who grew up on the 100-acre [40-hectare] farm, playing hide and seek in the barn, doing chores, and skating on the pond in winter.

Construction of the log barns began in the 1870s by Peter O'Neil, who took the 100 acres [40 hectares] of land out of the Crown. The barns are situated next to a creek that had a bounty of large pines and cedars along its watercourse. Over time the O'Neils pieced together six log structures in what became two main barns.

What magnificent buildings they were! The main barn, still standing, measures 64 feet by 25 feet [19.5 m x 7.6 m]. (See photo 7) The logs are 32 feet [10 m] on the long side, with a double wall running the width of the barn. The other barn, called the "long one", now all but collapsed into a tangle of wood, once measured an incredible 96 feet by 25 feet [29 m x 7.6 m]! (See photo 9)

4-2 *Ray McClocklin at the entrance to the main barn in the 1950s. (courtesy of Marg Martin)*

4-3 *The main log barn is seen behind the kids playing in this photo taken in the 1940s. (courtesy of Marg Martin)*

4-4 *Mrs. McClocklin and one of the children during the 1940s. The famous dinner bell is seen just at the top of the photo. (courtesy of Marg Martin)*

4-5 *A great shot of the main barn and driving shed and, as usual, a child playing in the foreground. (courtesy of Marg Martin)*

4-6 *The log barns figure into many family shots including this touching photo of Verna and her future husband Walter during the 1950s. The barn on the left is the first of five barns, all attached to each other, that headed down to the creek but have since collapsed. (courtesy of Marg Martin)*

At some time, O'Neil must have made a decision to keep the log barns instead of constructing the timber-frame buildings that must have been popping up all around the neighbourhood. "He just kept building on to it," said next-door neighbour Franklin Jackson, who has continued the tradition of farming the land his great-grandparents took out of the Crown in the 1870s. At one time, Jackson recalls, O'Neil had a steam-powered threshing machine that threshed the grain of 110 farms in the area. It took O'Neil and two hired hands just to run the steam engine, while the farmer and neighbours pitched in to thresh the grain.

Inside the long barn the second storey was used as a hay and straw mow and was further supported by a timber frame. This barn was a series of three structures that were eventually put under one roof. The first log and timber-frame structure in the long barn was an implement shed with a root cellar for potatoes under the main floor. The second building was the granary where the threshing was done and straw blown into the mow above it. A ramp was constructed from the second-floor mow so that hay and straw could be taken

4-7 *The main barn still standing in 2000. The doorway is the entrance to the stables and slightly above it is the opening where the hay was put into the mow.*

down to the main barn to feed and house the livestock. "It was steep and tricky with forkfuls of straw," remembers Marg, Verna's sister. The last section of the long barn housed young cattle. An opening provided access to a yard and the creek to drink.

What neighbours remembered most was a large cast-iron bell on top of the roof. It could be heard for miles around, summoning Ray McClocklin from the fields for dinner.

Most of the animals were housed on the ground floor of the main barn. Marg remembers the daily chores that had to be done by members of the family. "We never really looked at it like work but something that was part of our life," she said. Besides hand milking the cows—all of them named—every day, chores included churning butter, washing the cream separator and pails, bedding the animals, and collecting eggs from chickens and ducks.

"A good place to hide when mom was angry was in the pig pen. She never looked in there," laughed Marg. She remembers that her father once brought sheep to the farm but when a ram went after her sister, Verna, their father shipped all the sheep and never kept them again. Horses were stabled in the south half of the main barn and the straw stored above in the mow. The only way to fill that mow was through an opening on the south side of the barn. "As the straw was blown up there by the thresher we would carry it to the other end of the barn [60 feet or 18 m], until it was filled up," said Verna. "Now, that was hard work." (See photo 8.)

4-8 *The south side of the barn showing one of the livestock entranceways. Above this doorway is an opening where the straw from the thresher was blown into the mow.*

4-9 *The five log barns under one roof leading down to the creek have all fallen down. The bottom logs of the buildings decayed and, with a heavy snow load on the roofs, the log barns collapsed.*

Because the barns were built without foundations, settling began to occur, evidenced by the large gaps between some of the logs. The weak link in log barns was the bottom layer of logs, sitting on the ground or on a bit of stone support. These logs would invariably, over time, succumb to rot from exposure to rain and snow. So it was with the fallen-down long barn. The bottom layers were mostly rotten; the weight of the rest of the logs above them, including the mow and the roof, made the building collapse. The main barn has been saved by the dry ground it sits on and the strength of the saddle-notched corners.

If you look closely at the ends of the logs, some show marks of being cut with an axe and some with a saw. Why would this be? I deduced that the cedars were felled with an axe and those are the marks we see on some of the ends. The rest of the tree was cut up with a bucksaw that left the markings we see on the others. (See photo 10)

Reminiscing always seems to include stories about the animals that formed such a central part of life on the farm. The McClocklin sisters remember "Danny", the only horse born on that farm. "We totally spoiled him. I remember Father holding Danny's head and saying to him 'Tell me a secret, Danny' as if the horse could talk," laughed Marg.

Marg averted one near tragedy in the main barn. A Coleman lantern had been tipped over in the barn during chore time. Skinny, twelve-year-old Marg, hearing the cries of "Fire!" miraculously managed to carry from the house a large copper boiler filled with water from that day's laundry. She trudged with that heavy boiler, water splashing over its sides, along the little path into the barn and put out the fire. "I don't know how I carried it," she said.

A beef ring was established at the McClocklin barn, over the granary. Each neighbour who was part of the ring would have a turn at bringing in a cow for butchering and each would then take a share of the meat. This happened once a month. Tabs were kept on who brought an animal in and how much it weighed. Verna remembers little slips of paper kept above the granary with the names of neighbours and dates.

The family lived through the depression years of the 1930s and scarce years of the 1940s. "We were a big family and poor, and little treats were considered big ones. One thing about all those years, we always had food on the table and we were never hungry," recalls Marg.

The McClocklin family photos depict children growing up, having boyfriends, getting married, father and mother working hard, all centered on the barn, the source of income. Marg summed it up, "We never looked at it as work but as a way of life."

4-10 *The main barn was actually two structures built next to each other with a maximum log length of 32 feet [10 m] to the centre. This view shows the meeting of the two. Shifting and shrinking of the logs has caused much of the chinking to fall out.*

5. First Quota Chicken Barn

When the railway came to Owen Sound in 1873, Thomas Willsie decided to build his farmstead next to the rail line, rather than the mud trail that sometimes was called a road. This little trail leading into Owen Sound we call today Highway 6 & 10.

The railway line angled through the 50 acres [20 hectares] and was about a kilometre from the road. This was an important factor later in the farm's history.

The railway provided easy and convenient transportation for Thomas and his family. The train would stop at their farmstead and they were able to travel to Owen Sound in 15 minutes, where a wagon ride might take an hour or more, especially coming back south up the very steep rutted trail. At that time all of Owen Sound was located down by the harbour, and up on the hills above the town the land was dotted with farms.

Thomas built a barn by the tracks along with a fine stone house that is used today as a storage shed. In 1882 Archibald Mackay bought this farm and the parcel next to it (two parcels of 50 acres or 20 hectares). Whether the train wouldn't stop any more or because the road improved, we'll never know, but Mackay decided to move the barn closer to the road.

When the two side by side farms were bought, there already existed on the south farm a house and barn. The barn that was moved was joined with the other to make one very large structure. The original barn was 40 feet by 40 feet [12 m x 12 m] and its new cousin was 44 feet by 54 ft [13 m x 16 m] making the total length 84 feet [25 m]. These two barns would

5-1 & 5-2 *Ken MacLeod's barns, just before the roof was dismantled, being taken down in April 2000. The beams, largely hardwood, will be sawn into boards and probably sold to woodworkers. (courtesy of Ken MacLeod)*

5-3 *The barn in the foreground was moved. It originally stood at the back of the farm, built next to the railway tracks. Note that under the ramp is a root cellar, a common construction detail in Queen's Bush barns.*

eventually turn into four floors housing one of the first chicken quota operations in Ontario.

Ken MacLeod, who bought the farm in 1946 for $7,300, remembers Highways 6 & 10 as still full of pot holes in winter. "Mostly it was the cutters, whose tracks were wound up with chains to keep them from slipping, that would gouge out the road," said Ken.

In 1946 there was no electrical power or good water on the farm. After electricity came in the early fifties, Ken had contractors drill five different holes looking for water without sulphur or iron in it, but never found any that was suitable. In 1975, at a cost of $7,000, the farm was hooked up to Owen Sound's water system. "There just isn't any good water up here on the hill," said Ken.

Ken's first farm business venture was raising mink for fur. "Furs were a necessity when women were walking or driving in buggies in winter," said Ken. However, after the second World War, as cars became more commonly used, the market for fur changed. With the popularity of the car as the main mode of transportation, the demand for fur went down drastically.

One day in the early 1950s, a feed dealer told Ken to go into the chicken raising business. When Ken told him he didn't know anything about chickens, the dealer said, "You don't have to know anything, but it'll make you some money."

Ken not only took up this idea in 1952 but remained in the chicken broiler business for the next 46 years. Along the way he was instrumental in introducing the chicken quota system to farmers in Ontario.

In 1965, Ken sat on the first board of the Chicken Producers Association of Ontario, and attended a subsequent 68 meetings. "Supply management works for the farmer," said Ken, who was an outspoken advocate of selling the idea to the rest of Ontario chicken producers. "It was a tough time but it was a chance for farmers to make a decent living," he added.

Two more floors were added to both barns and by the late seventies, Ken had the maximum quota allowable, 30,000 broilers. Insulation was added to keep the chickens warm, yet the barns were never totally suited to broiler raising. Feed had to be carried up

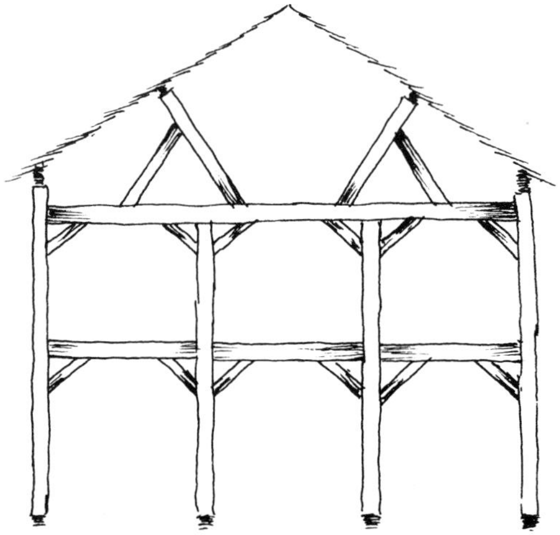

and down the stairs; water and plumbing always had the potential of freezing in winter. A modern one-storey barn was built in 1990 and a few years later the double barns were closed to housing chickens.

"The problem today with the quota system is that the big guys have figured out the loopholes," Ken said. He lamented the fact that large corporations had circumvented the 30,000 quota limit by buying more quota in their employees' names, gathering huge amounts of chickens and quota. "The system was supposed to help the farmer but the Chicken Producers Association never had the guts to properly police the system," he added.

The Queen's Bush has produced many outspoken, colourful people, and we can count Ken MacLeod as one of them.

The two barns perfectly portray the two kinds of post and beam methods that prevailed at the end of the 1800s. The front barn uses its long middle posts to support the purlins which keep up the rafters and roof while the back one brings in the engineering power of

5-4 *An excellent comparison of the two most common barn framing methods in the Queen's Bush. The barn on the left uses the queen-post method with braces rising from each of the bents to support the roof (already removed in this photo). The one on the right has high centre posts in each bent that directly support the roof through the purlins.*

triangles by using long braces of the queen truss system in its support of the roof. Both systems were very common, and usually the middle tall post system depended on the availability of long-enough and big-enough trees to make those beams. Most of these posts would need to be longer than 24 feet [7 m], and a minimum of 10 inches by 10 inches [25 cm x 25 cm] if they were hardwood.

In early 2000, Ken decided he did not need the barns any more and sold them to a local contractor in the business of taking down barns. The 100-year-old maple, white ash, and pine beams would be sawn into planks that already have a beautiful patina. Most would be made into furniture.

Although some barns in the Queen's Bush are seeing the demolitioner's hammer, many more have survived one hundred years of storms, tornadoes, snow and rain and today still hold the farmer's harvest in the mow and livestock in the stables. They are a credit to the building expertise of the pioneers, who have created a long-lasting legacy in timber-frame barns.

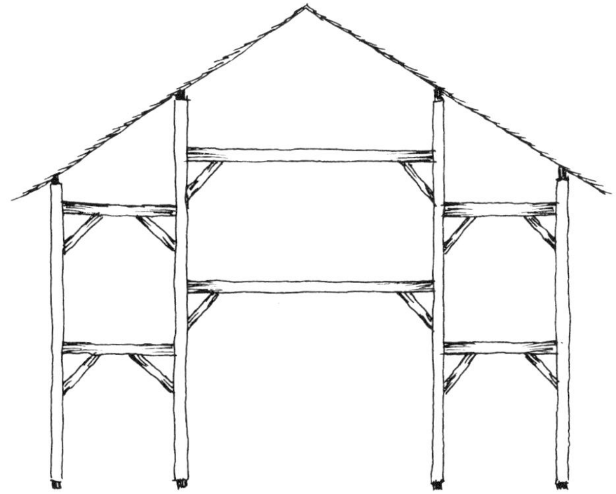

5-5 *The sun shines through the timber-frame structure for the last time. The outside posts have foam residue left over from the period when the walls of the barn were insulated to keep chickens warm.*

6. The U-Shaped Barn

This set of barns has all the intrigue of a mystery novel—murder, death by accident, and the beauty and enduring attraction of fine craftsmanship.

"At one time this was considered the best farm in Grey County," said Harvey Ruhl, who spent 60 years of his life on this farm in Derby Township, the first north of Keady on the east side of the road.

6-1 *Air photo of Ruhl farm in the 1960s. The 112-foot [34 m] barn is most visible. This concrete silo has been replaced by a much larger one. (courtesy of Harvey and Grace Ruhl)*

One of the first owners of the 400 acres [162 hectares] of rolling land was a Mr. Duncan, who purchased the land in 1856. He built the original log house, which is long gone now, and the first barn, a 26-foot by 46-foot [8 m x 14 m] timber-frame structure in what was to become a complex set of three barns, forming a distinctive *U* around a central courtyard. Alas, after Duncan's initial accomplishments, he decided to go back to Scotland on business and was murdered on the return boat trip. It was said he was killed and robbed for the $12,000 in gold he was carrying. Was the money meant for a further expansion of his farm enterprise—more land, bigger barns? We'll never know.

In the late 1890s the farm had expanded enough that another structure was needed for hay storage. The main barn was then built, parallel to the original one but separated by 45 feet [14 m]. (See photos 2, 4, and 7) The grand, 7-bent, 32-foot by 112-foot [10 m x 34 m] structure was struck by tragedy on the day of the raising when one of the men working there fell to his death. When Harvey took me for a tour of the barns he pointed to the spot where the man fell: first bent on the south side, where the purlins met the bent plates. Perhaps the man was in a rush, because competition was fierce among the crew to finish their jobs first and the man fell from 20 feet, [6 m] trying to hurry.

The floorboards of this barn were made of tapered hemlock. Harvey explained it was a good method of saving wood. Since the trees tapered naturally at the ends, floorboards were cut 8 inches [20 cm] at the wide end down to 6 inches [15 cm] at the small end. The boards were then laid alternately small and large end to make a 14-inch [35 cm] board. This also made the floor a stronger unit.

In the early twentieth century the two barns were joined by a 60-foot by 80-foot [18 m x 24 m] barn, forming the *U* one can see today. With the completion of the third barn the combined length of the three totalled an amazing 238 feet [72.5 m]! (See photo 3.) Horses were able to enter through the middle barn's wagon doors and exit through the west barn's doors, a real luxury compared to having to turn around horses, or tractors today, inside a barn.

The middle barn roof was eventually raised to accommodate more feed and hay, and one can still see where the old roof had been on the inside, because the barn boards of the two barns where they joined were never cut off. As the last barn became the centre of operations, it took on the duties of the threshing floor and granary. Under the floor of this barn were 12-inch by 18-inch [30 cm x 46 cm] elm, some 46 feet [14 m] long, holding up the sleepers and floor.

Wood for barn construction varied according to what was available out of the bush on the farm. All the beams were from hardwood stock, including white ash, maple, elm, and beech. Mud sills were mainly elm.

6-2 *A close-up from the 1940s shows the second barn, which was built in the late 1890s. The narrow 7-bent structure was initially used for straw. (courtesy of Harvey and Grace Ruhl)*

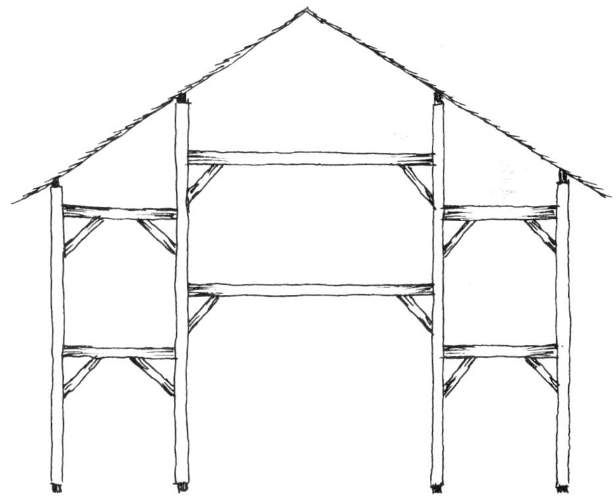

6-3 *The three barns total 238 feet [72.5 m] in length. The one on the right is the first barn, built in the late 1850s; the barn on the left, the second in the late 1890s. The connecting barn was built in the early 1900s and later was raised another 12 feet [3.6 m].*

6-4 *Not much has changed in the last 100 years to this beautifully massive barn which measures 112 feet [34 m] in length.*

The purlins reveal the era when each barn was built. The first had round, mostly cedar poles, the second barn roof was held by purlins flat on two sides, while the last barn had squared ones, which had come from the saw mill.

Inside the last barn was a 32-foot [10 m] wooden ladder tied to one of the posts. Built 80 years ago, its rails were made from a tamarack pole split in half. The two halves were joined by ironwood rungs to make a ladder. Harvey told me his father bought the ladder from someone in the area who specialized in making that type of climbing tool. It was used extensively for climbing everything from roofs to silos.

In the vast stabling, there were Beatty tie-ups for milk cows, stalls for horses and pens for pigs. Harvey built a small addition for chickens. Water was pumped from the well, near the house, up to a cistern next to the barn and then distributed by pipes inside the barn. The cistern also fed water by gravity to the house. Electricity and a house pump did not come to this farm until 1945.

The Ruhls, who came from Neustadt, bought the farm in 1919 and Harvey took over in 1950. The Ruhls bought the first rubber-tired tractor in the area, a Massey Pacemaker, in 1938. As on many prosperous farms in those days, a root cellar was built in the stabling of the barn, turnips being a main feed crop for livestock. Concrete walls which once held the crop in the stables have been knocked down, but the rough edges still remain. Small window openings in the foundation were used to fill the root cellar, and then the turnips were distributed for feed throughout the stalls of animals.

In its day, this set of barns was at the forefront of modern and efficient farming, with all the latest in technology. Today, they are used mainly for a beef cow operation, with a free range system in the stables. Gone is the mixed farm operation, and with it the stalls, tie-ups and enclosures, all built with a complete, integrated operation in mind.

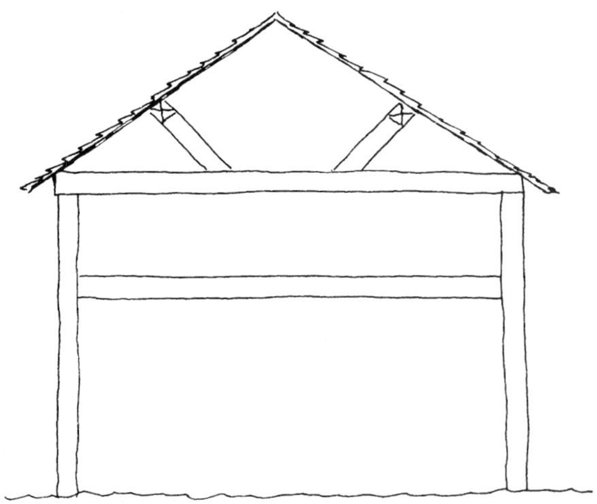

6-5 *A view of the inner courtyard shows the earliest barn with its stone foundation made to look like blocks. Barn boards put on 50 years ago have uneven ends—a building method that saved work.*

6-6 *Where two roofs meet, leaks may start; this point did so, until it was recently fixed. Here again, we see the unevenly cut barn boards.*

6-7

6-8

6-7 *The inside of the seven-bent barn has a peaceful permanence to it.*

6-8 *Looking at the inside of the middle barn from the long one. It's late fall and the mow is full.*

7. A Sullivan Hardwood Barn

After almost a century this barn has been taken down but the land is still used to produce abundant crops as in the past. And the past, on this farm, just north of Scone on the Grey-Bruce County line, is full to overflowing with the experiences of family farm life.

"Family life is pretty special to me," said Mel Turner, whose grandfather settled here in 1894, raised the barn in 1903 and in 1907 built a new brick house that is still standing.

A barn without its outer skin of barn boards is awe-inspiring. Laid bare is the intricate layout of the massive beams and the myriad of braces, small timbers, and ladders reaching up to the sky. This was a large barn by anyone's standards, measuring 64 feet by 74 feet [19.5 m x 22.5 m] and 50 feet [15 m] to the peak. There were five bents, with 34-foot [10 m] centre posts meeting the purlins which ran the length of the barn. (Photo 4) Double ramps led to two threshing doors which accommodated a bountiful harvest that sometimes overflowed the grain bins and took the hay, literally "to the rafters," said Harold Turner, one of four brothers and four sisters.

7-1 *An air photo of the 150- acre [60 ha] Turner farm. The fertile fields produced a variety of crops. (courtesy of Mel Turner)*

The hardwood frame was built by a local Sullivan barn framer named Shultz, who, Harold said, used an adze to square the great white ash, elm, and maple trees which came from the farm.

7-2 *An intricate web of timbers make up the barn with braces at each corner for strength. Beams on the outer edges of the barn always receive the most weather through the cracks in the barn board. Moisture travels through nails, causing the dark areas you can see on these timbers.*

Mel said it was the variety of work—planting, harvesting, cutting firewood, chores and just plain fun, all centred on the barn—that he remembers the most. The family milked by hand up to 20 cows during the 1950s. Behind the house a milk house was built, half buried in the sandy ground, for keeping the butter, cream, and vegetables. The cream was shipped to the Chesley creamery and the skimmed milk fed to hogs. One day lightning struck the litter carrier and gave Mel a good jolt, but what was most amazing was the pigs' tails all smoking from the strike. Luckily the barn did not catch on fire.

Meat roosters were raised above the pig stalls and fattened with chop soaked in warm water. Sheep were kept on pasture and the wool sheared and sold for cash. "I remember shearing sheep with hand shears," said Mel. "Now that was hard on your hands." There were also chickens and beef cattle to produce feed for and the 150-acre [60 hectare] farm proved it could grow a bounty.

The west side of the barn stables housed 12 horses, including two full working teams and two drivers. Horses were used for all the farm work, even towing the family car. During the winter months, the road north of Scone was not usually plowed and often the family car

7-3 *The barn as it looked for almost 100 years before it was dismantled by local Amish. (courtesy of Mel Turner)*

would be towed down to the Chesley road by horse. The same would happen on the return journey.

"We hoed five acres of turnips by hand," Mel said. I could tell by his eyes what a daunting task that was. The turnips were used as feed and sold to local stores for cash. "You couldn't sell oversize or undersize turnips. They had to be just right," added Mel. To be salable, turnips first had to be taken to a waxing plant in Dunkeld, and then stored in the barn root cellar.

Two types of corn were grown: feed and sweet corn. Each stalk and cob had to be cut with a hand-operated drum cutter. One person would crank the cutter and another one fed the stalks through until there was enough for the day's feed.

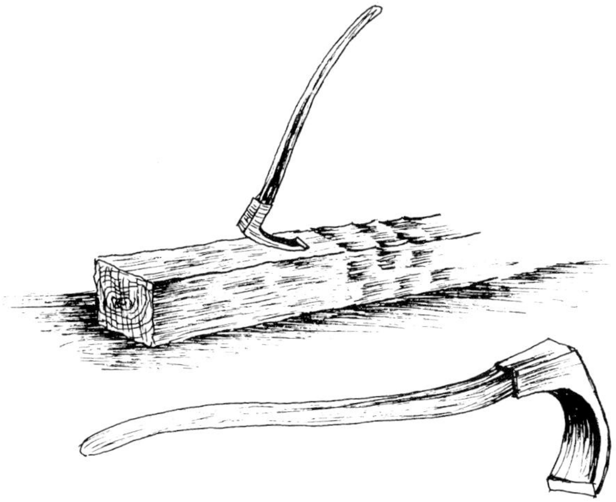

7-4

7-5

7-4 *Five bents and 34-foot [10 m] middle posts made for a large barn.*

7-5 *The full frame of the barn was notched in 1903 by a local Sullivan Township barn framer.*

7-6 *Ladders abound in this barn, to make it easier to climb to the top of loose hay piles which sometimes reached the rafters.*

Hay was cut by a horse-powered four-foot [1.2 m] mower. After it dried, the hay would be coiled by hand and forked onto the wagon. The hayfork hanging from the barn rafters transferred the loose hay into the mow, from where it was fed to the livestock during the winter months. "We would crawl up the ladder to the very top because the barn was so full," said Harold, "and then we would roll up the hay and throw it down for feeding (Photo 6)."

"We always had a good surplus," he added. On occasion they would sell some of the loose hay by delivering it on a wagon sleigh in winter but Mel said, "you couldn't get much on there, really, to make it worthwhile."

Threshing days were special. As the great steam-powered threshers lumbered into the yard, young Mel and his brothers would run out and beg the owner to blow the whistle. Later, when they got old enough to help with threshing their own grain and that of their neighbours, dinner was the highlight of the day. "We would eat like horses," Harold recalled. "There would be pies all over the place, but it was a lot of work for our mother," added Mel. About two cords of cedar, taken from the farm's swamp, would be supplied for the wood-fired steam engine. "All the neighbours would help one another and you gave a lot if you had it," Harold said.

To avoid trampling the outer round of wheat with the reaper binder, Mel's father would cut the grain with a hand scythe and cradle, while the boys tied the sheaves in bundles. The wheat would be taken into the barn to dry, threshed, and some of it ground for the family's flour supply.

Life was not always a bed of roses on the farm. Mel has never forgotten an incident that happened when he was five. His mother set off across the fields to the woods where raspberries grew. With one hand she pushed a baby carriage loaded with two of her youngest children and with the other helped along a toddler. She started picking the soft fruit but in a freak accident cut her leg artery on one of the canes. Blood poured out as she quickly tore a strip of cloth and bound her leg. Imagine her walk back up through the hot summer sun towing three little children, and what young Mel must have felt when he caught sight of her. "I'd never seen so much blood," said Mel. Fortunately help came quickly.

Chasing cows and running back and forth to school every day for lunch made Mel an excellent cross-country runner. "I would think nothing of running beside the buggy the five miles to my cousin's place," he recalled. Winning numerous local school prizes for victories in racing, Mel made it to the all-Ontario meet held in Toronto in 1958. He finished second, not bad for a country boy with no organized training, just used to chasing cows.

"We were always taught not to be mean-spirited, but we did do some innocent tricks around Hallowe'en," Mel said. Once the four brothers noticed a bunch of logs piled at a neighbour's gate. Hallowe'en night they moved the whole pile right across the driveway. Another time they rolled a very heavy flat stone onto someone else's driveway. "We had a lot of fun doing stuff," added Mel.

The barn was sold and dismantled by local Amish, then reconstructed into two barns on two separate farms in 1999, so it lives on to serve the needs of two more families. With it went some of Mel's best childhood memories.

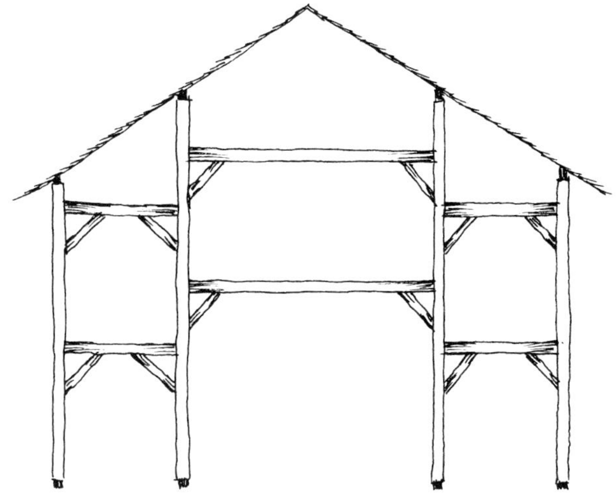

7-7 *A pastoral scene with the Turner barn situated on a rise and clearly visible from several fields away. (courtesy of Mel Turner)*

7-8 *The end of the line. Ladders beginning nowhere and ending nowhere.*

8. The Biggest Barn in Grey County

This barn is so large and tall that on a clear day, it is said you can see Lake Huron from the top of the roof. Surprising, since the barn is located in Normanby Township, 36 miles [58 km] inland!

Originally built in 1886, by the prominent Blyth family, the barn had two additions by 1902, one on the north side and one on the south.

There is rich rolling land on this 225-acre [92 hectare] farm, which was initially taken out of the Crown by John Blyth in the 1860s and stayed in the family for more than 100 years. John rose to prominence when he won the South Grey riding for the provincial Conservatives in 1879 and sat in the legislature for two parliaments. Unfortunately, he died suddenly in 1895, at the age of 45, of complications from appendicitis. He left nine children and his wife, Jane, who never fully recovered from his death and died eight years later.

8-1 *How the Blyth barn looked in 1903 with an addition at each side. The windows at the top were kept and installed again during the second raising. (courtesy of Howard Weltz)*

Eventually, his sons James and John Jr. took over the farm and in 1911, the barn was again reshaped, taking on the form we see today. And what a change!

The additions were incorporated into one main barn, while adding another 20 feet [6 m] to the overall length and raising the walls and roof over 15 feet [4.6 m]. To do this, the roof and its supports were taken down and another timber frame was assembled and raised on top of the original frame, a difficult but not uncommon practice. [See barns 1 and 20]. In fact, the builder was chided for building the frame six feet [2 m] higher than what the plans called for, to a total height of just over 50 feet [15 metres]. But what a grand frame it turned out to be. To support the massive new roof, three sets of purlins were installed, running the length of the building on each side of the roof. (See photo 4)

There is an unusual timber-frame feature in a beam that was notched at a forty-five degree angle half way up the side wall to the top of the

8-2 *The width of the barn stayed the same as in the 1903 photo, but the roof line, height and length have changed since a massive renovation in 1911. One of the largest barn frames built in Grey County, it measures 72 feet by 94 feet [22 m x 29 m].*

8-3 *The barn roof looms an amazing 50 feet [15 m] above the ground.*

8-4 *The immense six-bent timber frame has three sets of purlins running the length of the barn, needed to hold up the immense roof.*

first inside bent post. (See photo 5) This part of the new wall added on top of the old one in 1911 shows a skilled approach using the strengths of triangles to tie in heavy structures. Two more bents were added on the west side making the total length of the building 94 feet [29 m]. Looking up inside the structure, one sees a labyrinth of interlocking timbers.

James Blyth enlarged the barn because of his growing farming establishment. He specialized in purebred shorthorn cattle, but also kept pigs in the smaller straw barn stabling which was attached on the south side of the main barn in the late nineteenth century.

"I remember pulling and hauling turnips to the barn," said James Peter, who worked as a hired hand for the Blyths during the 1950s and '60s. "There were about six acres of them and harvesting by hand in November was a cold job, sometimes with snow on the ground," he added. The turnips were stored in the back part of the stables, which was converted into a root cellar the length of the barn and about 10 feet [3 m] wide. The cellar worked so well that it was not until the 1970s that the first silo was built on the farm. To make the turnips into pig and cattle feed each globe was put through a pulper. Turning a large handle moved

8-5 *An unusual feature of the barn frame is this 6 inch by 8 inch [15 cm x 20 cm] brace. The diagonal brace is notched into the wall where the old roof line began and rises to the top of the first interior post.*

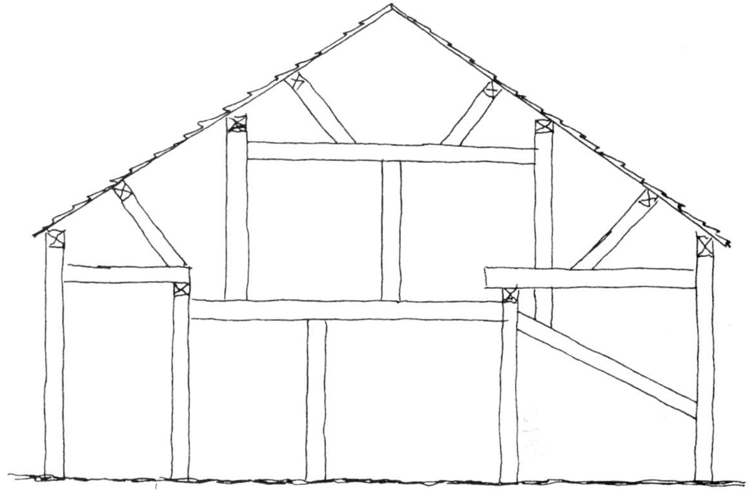

the blades on a drum cutting the turnips into chewable pieces. It's a job he remembered as hard manual labour.

"The dirtiest job was threshing the peas," James Peter recalls. The crop would first be cut by a mower outfitted with a special attachment that rolled the peas and stems into a bundle. The bundles would be thrown onto a wagon and taken to the barn to thresh. For a few years the threshing was done inside the barn, but it was so dusty and dirty that in later years the peas were threshed in the field. Peas were valuable as feed for cattle and mixed with oats for pigs, while the pea straw was used for bedding.

"There was so much grain grown it took three days of threshing to finish it," James Peter said. The sheaves of grain were brought into the barn, which had two main wagon doors on the north side. (See photo 8) James recalls the east entrance was used to store the sheaves until the threshing crew came. "The threshing floor and mow were so full they were filled to the top beam of the barn," about 35 feet [11 m] high! The Blyths' grain crop was always one of the last in the neighbourhood to get threshed as it was such a large job.

Barn threshing began at 7:00 a.m. and ended by 6:00 p.m. There was lunch at 9:30 in the morning with refreshments of sandwiches, cider, pie, cake, coffee, and tea. Dinner was served to hungry men at noon, followed by another lunch at three in the afternoon and a large supper at six. Some crews were threshing up to 50 barns during the season and lodgings and food were part of the deal. At each barn, neighbouring farmers came to pitch in and help. Those favours were returned when it was their turn for the threshing crew.

It was customary for neighbours to help each other with threshing and other farming endeavours, or if someone was sick or away. "I shouldn't say this," said James, "but people used to be more neighbourly."

The many tons of threshed grain were stored in a massive granary measuring 35 feet by 42 feet [11 m x 13 m]. There are two main doors leading into the ten interior bins.

An innovative stable floor was installed in the early twentieth century using short cedar blocks standing on end and filled between with sand. It was more durable than an earthen floor and it provided for good drainage of liquids, but cleaning out the stalls and aisles was another affair. "It was hard work. The shovel would hit the tops of the blocks because they were uneven. I wouldn't want to do it again," said James.

8-6 *Reshaping the barn in 1911 created a complex bent configuration with some unique notching.*

8-7 *The south side of the barn and loafing yard. There was a forebay on this side until the 1970s, when the interior foundation was taken out and replaced on the outside by a concrete one with openings wide enough for tractors. Note the straw barn at the far end, built in the late nineteenth century.*

The source of water for the house and barn was a spring from which water was pumped by a ram installed in the late nineteenth century and used until 1971. The water travelled to a cistern on the cold northeast corner of the barn. After winter set in, ice was sawn out of local lakes by the farmer and stored in a small framed shed built on top of the cistern. The inside walls were lined with a layer of ship-lapped boards to keep the heat from coming in while the cold water cistern below also helped to keep the shed cool.

Raising barn structures and taking them down can be dangerous jobs. James Peter almost lost his life when taking down a barn near Dornoch during the late forties. "We were lowering the first bent when the rest came down." He was struck by one of the beams and spent many months at the old General and Marine Hospital in Owen Sound.

The Blyth farm was sold in 1970 to Howard Weltz who continued in the tradition of raising beef cattle, trading the shorthorns for some Hereford crosses and farming the good land until he too sold it in 1999.

8-8 *There was plenty of room for two main doors on this massive barn. The doors on the left were used to stack the sheaves of grain and those on the right for threshing, with the granary right beside them. The small building attached on the left was the ice house, with the cistern underneath it.*

9. Double Barn

In the beginning of the twentieth century, skilled men with the right tools found work moving timber-frame barns. Once the pioneer settlement of the Queen's Bush was complete, farmers began to look to expansion. Fifty-acre grants of land were not quite enough to make a living from, especially with children who were growing up and looking for a farm future. Many farmers bought the farm next door, as the population shuffled between making a living from the farm and getting a job in one of the many manufacturing industries in the Queen's Bush.

9-1 *About 1905 Henry and Louisa Monk stand by an apple tree with only a single barn and attached straw shed seen behind them and the original house. The child, Alma, was mother to the present owner, Robert Engel. (courtesy of Robert Engel)*

Around 1908, Henry Monk doubled his Brant Township acreage by buying the farm next to his on the west side. It was good, productive, flat land and, with the new acquisition came a house and a barn. The small log house was left to settle with nature, but the barn was a great asset. For a while, chores were done at both barns, which were about 350 feet [107 m] apart. But, by the early 1920s Henry decided to move the 35-foot by 55-foot [11 m x 17 m] barn next to the one at the home place to make his work easier.

Neighbour Wilbert Musehl, now 90, who remembers doing chores in both barns when they were still apart, ran from school every day to join in the excitement of watching the barn being moved. "There it was, a barn in the middle of the field with one horse in front moving it," Wilbert said. Many other people came to watch also, lining up along the sides of the fields and giving their views on the merits of the move. "It was great entertainment," said Wilbert. Even though the land was fairly flat between the two sites, and fences were taken down, it still took a week to complete the move.

The contractor by the name of Eby from Elmwood hired to relocate the timber-frame structure, came prepared with all the necessary tools—jackscrews, rollers, lumber, pulleys, a winch—as well as four men, and a Clydesdale horse. "The hardest part was moving it off its foundation and down onto the ground," recalled Wilbert.

First the barn was lifted with 80-ton [73,000 kg] jackscrews from its foundation and then with maple logs inserted under it, slowly moved off. "It was huge. You'd think of it as impossible," Wilbert declared. "Yet, only one horse was needed to move such a great load to its new location 20 rods [100 m] away." A winch was used, geared down so low that when the horse pulled one end of the cable, "you could hardly see the barn moving," said Wilbert. Maple and elm rollers were placed under the barn and, as they popped out at the end, another horse took the rollers to the front again.

To prepare the site for the new barn, an addition used as a straw shed was first taken down. Since the straw shed foundation was not wide enough for the new barn, elm posts were set in place every 10 feet [3 m] from where the stone foundation ended. Finally the barn was brought to its destination, jacked up again and rolled very slowly onto its new foundation.

When matched, the new barn was two feet [60 cm] shorter widthwise, and about 5 feet [1.5 m] shorter lengthwise. But surprisingly, the threshing floors and two middle posts which defined that space, were a perfect match. As well, the heights of the horizontal girts were the same. Both frames were of the queen-post type, and many of the notching features, such as brace sizes, built-in ladder positions, and granaries were very similar, almost identical. In other words, it was most likely that the barns were built by the same framer, as both were constructed during the 1880s.

There they are, two barns next to each other. Looks odd, as most barn additions make an *L* or a *T* or are just added onto a gable end, making the original barn longer. The most difficult part in this match-up was where the two roofs met in a *V*. A new roof joining the two created a peak which shed the water and snow nicely. Henry had acquired another

9-2 *The main doors of the original barn lead onto the threshing floor. The middle posts lined up well with the new barn, which was probably built by the same framer.*

9-3 *Alma Monk and another woman, dressed in distinctive 1920s fashions, stand in front of the just-completed double barn. (courtesy of Robert Engel)*

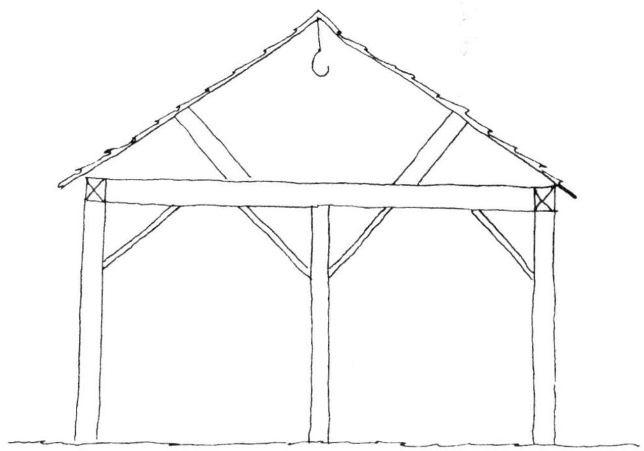

9-4 *A good view of the recently joined barn in the late 1920s with an uncle of Robert Engel showing off a newly acquired horse. (courtesy of Robert Engel)*

9-5 *Where the posts of the two barns meet are clamps installed to tie the two together. The bents for both barns were of the queen-post style and matched well.*

barn and he made the best possible use of it. It doubled the stabling and the volume of the mow, and here the barns sit, almost 80 years later, still fulfilling the needs of the present-day farmer.

When the post foundations began to settle in the 1960s, large clamps were put around the posts where the two barns met in the middle. Eventually, Robert Engel, the present owner, took out the posts and poured a concrete foundation. Then he removed the old inside stone foundation and replaced it with I-beams and metal posts in 1975.

I conducted many interviews in farm kitchens, where people were always very welcoming, and the Engels were no exception. With Robert and Wilbert, the conversation turned, as

9-6 *The barns in 2000 are a modern-looking enterprise with a large stable set up for 30 milking cows. The barn on the left was moved and joined to the one on the right in the 1920s.*

9-7 *Louisa and daughter Alma in the 1930s next to the new house built in 1921 at a cost of $3,000 with the double barn in background. (courtesy of Robert Engel)*

many others had, to threshing machines and early tractors. The light went on in Robert's eyes when he talked about his first combine and tractor. "It was a Mildmay [thresher]—they were small, cheap and versatile," he said. "And, my first tractor was an 8N [Ford]. Everyone had one." In 1953 Robert bought the used 25-horsepower tractor for $1,200 from a dealer in Chesley on a shared basis with his neighbour.

Until the late 1950s, the farm was a mixed operation, with chickens, milking cows, and pigs, all providing some cash and food for the family. In 1963 Robert renovated the stables into a dairy operation and milked cows for the next 35 years until ill-health forced him out in 1998, when he sold his dairy quota.

10-1 *The original owner, Joseph Neelands, registered in the 1871 census, seen here in front of the barn in 1895, with two horses, Prince and Dick. (courtesy of Wilmer and Mabel Beckett)*

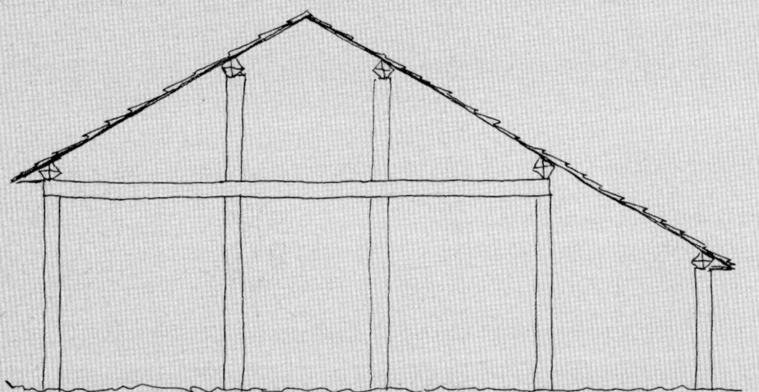

10. Salt Box Barn

The narrow dirt road that went by this farm was close enough to Owen Sound to become the important thoroughfare it is today, the Highway 6 & 10 by-pass around the scenic city. One hundred years ago there was a maple tree beside the driveway to the house. As time hurried along a few more were added and today these century-old trees almost obscure the view of the house and barn.

Photographs offer a rare opportunity to see this farm at 50-year intervals. The scenes take us from the boom of the late 1800s, with its community building bees and abundance of natural material, to the more mechanized, mixed enterprise of the 1950s and finally to today's specialized agri-business. With each era came a different set of buildings and silos, with different demands on the main barn and resulting additions.

By 1890, this Derby Township farm was in prime shape. A new house was built in 1873, a storage barn shortly afterward followed by a windmill and pump house, a driving shed and in 1890, the main barn.

The 60-foot by 60-foot [18 m x 18 m] saltbox barn reflects the type of trees available for building on this farm at that time. The frame is made from mixed hardwoods, with some softwoods used for the flooring. There are four, 28-foot [8.5 m] maple posts rising up around the threshing floor to support the roof. In the stable, two, 60-foot [18 m] rock elms support the main floor above, laid with boards cut from white pine. Some girts were shaped from white ash and some from beech. This was also the first barn in the area to have some of its beams cut at a sawmill. While all the main timbers have the distinctive marks from the broad axe, the smaller girts have been sawn, with the circular saw blade impressions left as evidence.

The main frame is 40 feet by 60 feet [12 m x 18 m] with a 20-foot [6 m] add-on built at the same time as the main frame. This gives the barn its distinctive saltbox shape, in this case, with a low roof to the south. The tall north side is built into the slope of a hill, giving the "bank barn" definition, and allowing the main barn doors to open onto the threshing floor and mow.

Inside the barn remains an original wagon rack lifter, installed to make unloading loose hay easier. A large wooden wheel was attached to a roller set between two of the longest posts on one side about 18 feet [5.5 m] up, with a smaller wheel and roller on the other side. These were used to lift the wagon racks and their loads by way of four ropes, one attached to each corner of the rack. The bigger wheel, which was ratcheted, would be turned by a rope and used as a pulley, raising the great loads of loose hay. The hay could then be pitch-forked off the wagon *down*—an important word here—as opposed to *up* onto the mow floor on either side of the threshing floor, making the whole job of unloading much easier. When the rocks were picked, the crops sown and harvested, and the final straw and grain stored inside, the wagon racks would be elevated high above the threshing floor and stored out of the way for the winter. (Photos 7 and 8)

In 1947, young Wilmer Beckett bought the farm from Roy Smith, and worked it for the next 52 years, selling it to the present owners in 1999. When Wilmer first purchased the farm, electrical power had been hooked up only a few years before. There was the original, distinctive Ontario storey-and-a-half house, driving shed and main barn. (Photo 3) A storage barn had been recently torn down and so too the windmill, with an electric pump replacing the wind's sometimes erratic power. The Neelands had taken the land from the Crown and Roy Smith, the second owner, had married into that family, making the Becketts just the third owners of the farm.

Due to a "quirk" in the survey, the farm was 110 acres [44 hectares] instead of the usual 100 acres [40 hectares]. The story goes that one day during the 1860s, when the land was being surveyed, the surveyor and his crew spent quite a bit of the day in the tavern at Kilsyth, by then a thriving little community, three kilometres west of the farm. When they marked out the lands for the future Neelands homestead, the survey ended up with 10 extra acres. We don't know if the surveyor realized he had made a mistake and decided to cover it up or no one noticed until the 1970s when the farm was surveyed again because Wilmer purchased the farm next door. Wilmer suggested, with a twinkle in his eye, that somewhere along the top of Derby Township, is most likely a farm with only 90 original acres [36 hectares].

10-2

10-3

10-4

10-2 *The house and barn as seen in 1904. Joseph Neelands and his family pose in front of the house playing croquet. The image is of a prosperous farm and a gentleman farmer. The one maple tree is still standing and a dozen more have been added. (courtesy of Wilmer and Mabel Beckett)*

10-3 *An air photo, taken in 1953, shows a steel roof on the barn, a driving shed added to the farm yard and a concrete silo in front of the barn. The house remains virtually the same as it was fifty years before. (courtesy of Wilmer and Mabel Beckett)*

10-4 *The year 2000 sees a big change in the house after a fire in 1983. The concrete silo is gone, replaced by two Harvestores. The barn is still very much the same but is painted black.*

10-5 *Forty-two feet to the peak from the mow floor, the roof is still supported by round rafters. The door on the right led to a fairly large chicken operation during the 1950s.*

10-6 *The barn was originally built in 1890, with a high north side and low south wall.*

When Wilmer took over the farm he continued with a mixed operation which included dairy cows, pigs, chickens and some beef. Threshing was still a community affair, with neighbours coming over for "threshing bees" to help put the grain and straw in the barn. The favour would be returned when it was the neighbours' turn to thresh his grain.

"They were the best times. The local wives would get together to make all kinds of good homemade food and pies for the crews. My, those men could eat," said Wilmer's wife and partner, Mabel.

Mabel reminisces about the strong community back then where people knew and helped each other. "Life is a lot faster and more complex, now," she said.

Meanwhile, Wilmer was making changes to the farm. The house was renovated after a fire in 1983. He took down the old cement silo built at the beginning of the twentieth century in front of the barn, and erected two Harvestore silos, one behind and one to the west of the barn. An addition was built onto the east side of the barn and a free-standing open-air feed building was also constructed. Fence lines were taken out and more land was cleared as Wilmer tried to increase the productivity of the farm by specializing in beef.

Salt Box Barn

Making a decent livelihood from a family farm in the last 50 years has become increasingly more difficult. "Right from the start we never made a complete living from the farm," said Mabel. "Things just changed and farming was never the same again." Gone were the "bees" of neighbourly help, diverse crops and livestock, each valued for its place in a complex agricultural enterprise. Look at that nineteenth-century photograph again. There is a beautiful simplicity to that era that is gone forever.

10-7

10-8

10-7 *and* **10-8** *The wagon rack lifter. The two photographs show the large and small wheels and shafts used to lift the wagon racks with their great loads of hay. Note that the large wheel is ratcheted.*

11. McClure's Mill

In pioneer days timber-frame construction was not limited to barns. Houses and stores were also built from large beams, but some of the most impressive timber-frame buildings were water-powered mills.

If you go to downtown Chesley and follow the North Saugeen River west for half a mile, you will come to the three-storey timber-frame structure of McClure's Mill. Built of 10-inch x 10-inch [25 cm x 25 cm] sawn hemlock, pine, and white ash timbers, the imposing structure harks back to the days when water power was king and settlements grew where a mill was located.

11-1 *The great structure of the mill rests peacefully next to the North Saugeen River. The first bridge was built by William Elliot to allow customers from Elderslie to get feed and flour from the mill.*

Mills were meant to endure weight and work, and timber-frame construction was a natural selection for this kind of early industry. Timber size and the kind of wood used for the construction of mills were important because the ever-flowing rivers were a power to be reckoned with. McClure's Mill was no exception. (See photo 5.)

On the river side the mill is supported by cedar logs with diameters of 4 feet [1.2 m], submerged in the river all these 120 years. The present owner, Bill McClure, great-grandson of the original builder, confirmed they are still sound. It seems that Bill's father, Jim, who took over the mill in 1923, was unsure of the condition of the cedar logs in the early 1980s. At the northeast corner, the place where the water first hits the mill, Jim decided to replace one of the cedar support pillars with cement. He closed off the water and after much jacking and putting in supports, the cedar log was pulled out. Much to his surprise, the log was still sound! Unfortunately, in the pulling out, the log had been chipped and damaged and the cement pillar had to be poured. The rest of the cedar logs were, of course, left in place to support the 50-foot [15 m] high timber-frame structure.

11-2

11-3

In 1883, William Elliot, already the builder of many dams and mills, including other mills in the area, saw the potential of developing waterpower where the mill now stands near Chesley. He was so ingenious around waterpower, and built so many buildings that people had nicknamed him "Beaver Bill".

To build the mill's flume, (the narrow channel of water diverted to turn the turbines), Beaver Bill let water erosion do part of the work for him. He began with a simple one-furrow plow. Starting upstream, next to the river's edge, he made a one-furrow pass by the mill where the four future turbines would sit, and continued in a wide arc, returning to the river about a half-kilometre distant. He then opened the river into his furrow, allowing the water's rush to wash away the heavy clay and expose the rocks below. When he blocked up the river, the stones that lay uncovered in the ditch were picked and the process repeated until a 12-foot [3.7 m] deep flume was built. So much for the need of modern high-hoes and bulldozers!

11-2 *Mill owner Bill McClure, right, and his son, Bob, in front of the mill.*

11-3 *Part of the workings of the mill on the third floor. Some of the gears may have come from a two-storey foundry built by William Elliot in Chesley in 1876.*

11-4

11-5

The dam was another ingenious construction. It was built of loose stones, with large boulders wedged into the bottom of the river. Even after the disastrous flood of 1916, which washed away concrete dams upstream and downstream, the Elliot dam stood up to the river and is still there.

In 1923 the McClure family bought the mill, and operated it as a grist mill. The basement was turned into a machine shop with a metal lathe and cylinder mill—all run by water power. The original four turbines, over the years, ran a chopper, a grinder, a roller, the elevators, and even a DC generator that charged 6-volt batteries during the 1940s and provided electricity for the house.

Flour milling was an important enterprise at the mill, with milling stones being installed at the end of the nineteenth century. Stones need regular cleaning as they

11-4 *The simple built-in hoist was used to lift the grinding stones. Wheat grain embedded in the stones would need to be chipped out to make the stones sharp again.*

11-5 *A double beam was used for the 36-foot [11 m] wide mill to carry the extra weight of the grain on the first and second floors. The top beam is made up of two hemlock timbers, each one 18 feet long, [5.5 m] meeting in the middle, while the bottom is a 14-foot [4 m] white ash, making the dimensions of the two 10 inches by 20 inches [25 cm x 51 cm]. The joists are on 20-inch [51 cm] centres.*

become embedded with crushed grain. To manage this, a simple permanent hoist was notched into an upper and lower beam and used to lift the stones. (See photo 4). Flour milling was discontinued in the 1960s and the circular stones were stored under the front platform of the mill, where they remain.

This timber structure is made mostly from hemlock and pine. Hemlock was in abundance in the nineteenth century and Beaver Bill probably got the logs at minimal cost. Most of timbers for the large barn frames in those days were squared up using an axe for scoring and a broad axe or an adze for hewing. Luckily for Beaver Bill, there was a sawmill in Scone, about three miles [5 km] away. He was charged $5 per thousand board feet. Besides the timbers needed for the frame of the building there were also the 3-inch by 10-inch [7.6 cm x 25 cm] hemlock joists, 3- inch [7.6 cm] pine and hemlock tongue and groove flooring, and 3-inch by 5-inch [7.6 cm x 13 cm] braces.

Donalda McClure remembers during the middle of the last century, long line-ups of sleighs outside the mill, waiting for their turn either to have their grain ground or to pick up pre-ground grain trucked in from Owen Sound. "We sometimes worked at night, but we had a DC generator for lights. One of the first buildings in Chesley with electricity!" said Donalda. It was always a family affair with up to four members working at the mill during its heyday.

Today, the water-power capabilities of the mill are not being used and the mill is selling an assortment of fertilizers, pre-ground feed, seed, and coal.

The future potential of the mill lies in converting the four idle turbines to produce electricity for the general market; perhaps as electrical costs continue to rise, McClure's mill will once again use its natural water power for another line of business.

12. Beaver Bill's Wood Foundation Barn

If you cross the little bridge which spans the North Saugeen River right next to McClure's Mill in Chesley and drive north up the winding hilly road, the first barn on the east side sits proudly overlooking the river and the mill. The disused railway line bisects the property at an angle, and played a key role in the construction of the barn.

12-1 *The Elliot barn with a silo built in the 1940s, facing south. The structure is being supported on this side by wooden posts.*

At first glance the small barn, measuring 30 feet by 46 feet [9 m x 14 m], looks like most in the area. However, the builder was none other than the innovative Bill Elliot, also known as Beaver Bill, who built McClure's Mill, about a stone's throw away. Probably Elliot lived here, building the barn and a homestead in the late 1870s, so he could have a place to dwell while he tackled the large project of the mill.

What is unusual about this barn structure is its foundation—it's made of wood. The barn is banked into a hill on the north side, with a solid row of railway ties standing upright, forming the foundation on the north and west sides of the barn. The railway ties are sunk into the earth for 12 inches [30 cm] and then banked up on the outside. A normal sill plate beam, 10 inches by 10 inches by 36 feet [25 cm x 25 cm x 11 m], in this case elm, meets the railway-tie foundation. On the south side, a wood post foundation is utilized, with a post sunk into the ground every 10 feet [3 m].

Beaver Bill again used his ingenuity to build a wood foundation and make it last, when others were laboriously building with stone. Since the railway line was only 50 feet [15 m] from the northeast corner of the barn, he must have been able to take any extra ties or buy them from the Grand Trunk Railroad as they were building the line at the same time Beaver Bill was building the barn. The first train reached Chesley in 1881.

12-2 *The barn is banked into a steep hill. Note the absence of a familiar stone foundation.*

12-3 *A row of standing railway ties forms part of the foundation on the west and north sides of the barn. Openings were simply cut out of the ties to form windows.*

12-4 *On the east side, the foundation is made from posts and horizontal planks to keep out the weather and soil. The wood has lasted quite well considering the barn was built in the 1870s.*

12-5 *At the top of the hill there is good access into the barn and threshing floor from the north side.*

The east side of the foundation is also wood but the barn sills are supported by posts with horizontal planks nailed on the outside of the posts to keep out the weather and dirt. These have held up amazingly well for being more than 125 years old.

Although the four-bent barn is not very large, 30 feet by 46 feet [9 m x 14 m], the timber-frame structure was well built. Double braces strengthen all outside connections and the variety of wood types used included white ash, pine, and hemlock.

Another oddity of the barn was the number of windows that were originally installed. On the south side five windows were set in and four on the north. There were three windows on the east and three on the west side of the foundation. Window holes were pre-cut out of the railway ties before the ties were set into the ground for a foundation.

Unfortunately, some of the ties have fallen over on the west side of the foundation but luckily the 30-foot [9 m] elm sill plate is still holding up the barn quite well. (See photo 3)

Bill McClure owns the barn and land around it now, but admitted the stable is too small to accommodate beef cattle, which he houses at another farm. The barn and the house, a brick building that has seen better times, are at the edge of Chesley, and when Beaver Bill originally bought the land, he had it divided into house lots anticipating Chesley's future expansion. There are still subdivision plans on record naming imaginary streets running past the barn and house. This farm is part of Elderslie Township, and Chesley never has grown this far west. Perhaps because the farm was "on the other side of the tracks," development seemed to stop just before its boundaries.

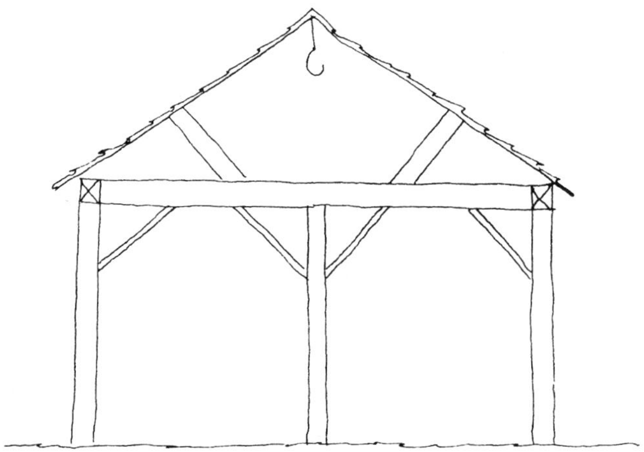

13. Four Barns Later

You never know where some farmers' stories will take you. This one leads us from Grey County and a burned barn to Alaska, and back again.

When Frank Meredith bought this Proton Township farm in the late 1930s, it came with a large house, what was left of a burned-down barn and 132 acres [53 hectares] of farmland. He made plans and built another barn but fortune was not smiling on this farm, for 20 years later, after Frank had sold the farm, this barn was also consumed by fire, the third one on that site.

The four photos on these pages are the only record left of this barn and house, for the house was torn down to make room for a new one.

This is the first farm in Proton Township, if you are coming from Dromore, on Egremont Township's Concession 18. No road separates the two townships, just an imaginary line where the boundary runs through small lakes and swamps north all the way to the outskirts of Priceville.

When Frank acquired the farm he was still a bachelor and had a job with a local construction crew, but what he needed most to start farming his new land was a barn. He hired Harold Boyle, a local builder, to construct for him a 40-foot by 76-foot [12 m x 23 m] timber-frame barn.

By the 1940s there were not many of the large trees left that could be used for barn building, so a surplus barn was found 6 kilometres away in Proton Township. This barn was taken down piece by piece, put on wagons and pulled by truck to Frank's barn site.

The foundation needed two new sides because Frank's barn was a little wider and 20 feet [6 m] longer than the one Harold had taken down. The concrete was turned in a mixer run by a small engine and then shovelled by hand into the forms. Years later, when Frank took down a part of the concrete foundation to make another door, he found it easy. "I could take most of the foundation out with a shovel. There was not enough cement in it," said Frank from his home in Dundalk. "Yet the foundation held for all the years I was there so I guess it was good enough," he added.

13-1 *One of Harold Boyle's crew, who built the barn in 1940, shinnies up the gin pole to attach the pulley and ropes which helped to secure the pole. (courtesy of County of Grey-Owen Sound Museum)*

New timbers were still needed because the added barn was longer. Frank recalls that Harold was in his bush "many a time" cutting mostly elm for new beams to extend the frame.

By 1940 neighbourhood barn raisings were getting to be a thing of the past, and many young men were away at war in Europe, so Harold Boyle with his crew of five raised the barn bents with the use of a gin pole and tractor. They chose a fine straight elm that would be at least 10 feet [3 m] higher than the top of the barn frame. Guy wires and rope secured the gin pole in the centre of the barn floor. Then, after each bent was re-assembled on the floor, it was raised by the use of a pulley fastened near the top of the gin pole and pulled up by a long rope attached to a tractor. Lighter beams connecting the bents were manhandled to their places by strong hands. Larger connectors could also be lifted up by the use of the gin pole. New oak pins were pounded into place.

It was long arduous work and it took the crew of five all summer to finish the job, and for that, Frank paid $350. "It was a lot of money in those days," Frank said.

"The farm was not in good shape when I bought it, but I improved the land and the value with every cow I bought," said Frank, who paid $14,000 for the 132 acres [53 hectares]. To help pay for some of the expenses, Frank, who had commercial road construction experience, agreed to an adventure of a

13-2 *The new barn frame, almost finished. The crew on the roof is nailing on strapping for the metal sheeting. (courtesy of County of Grey-Owen Sound Museum)*

13-3 *Left-over lumber outside the almost-completed barn. The crew is just finishing up with the stabling. Many windows in the foundation would make for bright stables. (courtesy of County of Grey-Owen Sound Museum)*

lifetime—a foreman's job on the construction of the Alaska Highway.

After selling his stock he took the train out west and arrived in the wilds of the north in 1942. "It was a little different than my farm," admits Frank. The construction of the Alaska Highway was deemed to be a war-time emergency because of the threat of attack from Japan and possibly Russia, across the narrow Bering Strait. A highway was considered essential to transport weapons and men.

Frank's job was to make sure that trucks and equipment were at the right place at the right time, whatever the weather. This consisted of moving around bulldozers, graders, trucks, "and a lot of men with shovels," said Frank. Transporting bulldozers meant they had to be taken apart, as they were too heavy for the trucks to haul them intact. The crews worked in the hot, summer mosquito season and through the long Alaskan winters. "The winter of '42 was one of the coldest. Bulldozers kept getting iced up and stuck in river beds because they were too heavy to cross over bridges. We had to dig them out of the snow and ice," Frank remembers.

13-4 *An aerial view shows the farm in 1949. Today the house and barn are both gone. New structures are in place and history can be recalled only through memories and old pictures. (courtesy of Frank Meredith)*

After two years Frank came back home to Egremont and married Edna. She began teaching at Swinton Park School, about six kilometres from their farm. The winter of 1947 was one of the harshest in recent memory, with record snowfalls and low temperatures. Edna especially remembers one day that winter, coming home from school in a cutter pulled by a horse. It had been another terribly stormy day, but she thought she could make it home. The snowbanks at the edges of the road were above her head and as she came around a corner she found herself up against another cutter, abandoned in the middle of the road. She could not get around it nor could she back up. With nightfall rapidly approaching and temperatures dipping, she unhitched her horse and found her way to the nearest farm, where she spent the night. Poor Frank, without any means of communication, didn't know what had happened to his new bride until she came back the next day.

13-5 *"All the neighbours got together to help thresh," recalled Frank Meredith. Scenes like this happened all over the Queen's Bush. (courtesy of Glen Stephens)*

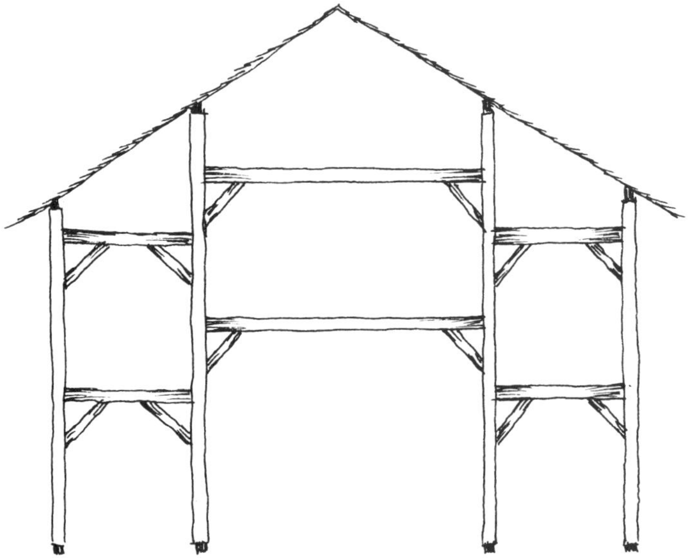

This barn was built large enough to house a double threshing floor. By late fall each year, a wood-fired steam powered tractor and threshing machine would arrive at Frank's barn. With the thresher inside the barn and the steam-powered tractor belching smoke outside, sheaves of grain that were stored on top of the loose hay and on the other threshing floor, were methodically fed into the thresher. "All the neighbours got together to help each other thresh," said Frank. After one farm's grain was finished, the threshing gang would move on to the next farm, and the adjoining neighbours would help out.

Water was always a problem on the farm. The old well had caved in and after a big effort Frank was able to reline it. He took off the tire from his first tractor, an Allis Chalmers B, and used the rim of the wheel to run a pump. "That was so much better than hauling water," said Edna. There was still not enough water, so they hired a well-pounding crew to get them a deeper well. The crew pounded for seven weeks looking for water. "I almost went out of my head over the noise, day in, day out, all summer long. They didn't even stop for lunch!" said Edna.

In 1952 Frank and Edna wanted a change and put the farm up for sale. To their amazement, it sold in two days and they searched frantically for another farm, finding one just south of Flesherton. Ten years later, the barn that Harold and the crew built for Frank burned to the ground, the third barn on that same site to go up in flames. Now another barn, a pole type, sits on the same spot, making it the fourth barn built on this Egremont farm.

14. Royston Park: A Gentleman's Farm

William Roy had grand ideas for his farm and estate near the edge of Owen Sound, in Sarawak Township.

He had a large two-and-half-storey stone house built in 1873, with a sweeping view east to Georgian Bay, which his 200-acre [80 hectare] farm bordered. The stone house, though, was meant only to house the servants. What he intended to build onto its front was a castle. There is still evidence of his palatial plan. On the east side of the house are the pins installed into the stone that were meant to connect one day with the stone walls of the castle. Although the grandiose plan never materialized, William Roy named his farm Royston Park, perhaps envisioning men practising skeet shooting and families in their finery coming to play croquet and picnic by the water's edge.

14-1 *There were two threshing doors to this grand barn with a centrally located granary.*

14-2

14-3

14-2 *The maple dowels located at the corners of the granary kept the wagons from hitting the sides of the structure. The horses could unload their straw and hay and keep going forward around the granary and out the other threshing doors.*

14-3 *Clay tile air vents were built into the foundation for better air circulation in the stables. The foundation and house were built from the plentiful amount of good construction stone available on the farm. It's one of the reasons the original owner had the grandiose idea of building a stone castle.*

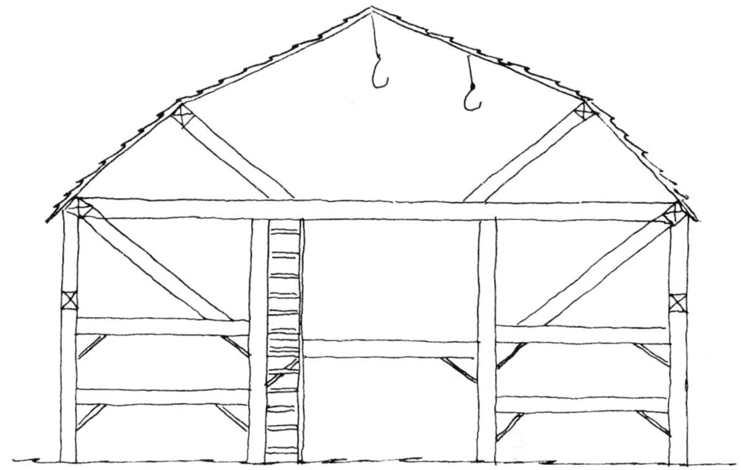

The 200 acres [80 hectares] was taken out of the Crown in 1863 by William; he built the first timber-frame barn 10 years later. He rose to prominence and became the first president of the General and Marine Hospital in Owen Sound. Originally from Scotland, he imported various hardwood and softwood trees from that country to beautify his farm-estate. An orchard was also planted, known as the Royston orchard, of 90 pear trees and an assorted selection of plums, apples, and gooseberries.

The builder of the present-day barn bought the farm in 1896. William Morrison Jr. had the unenviable experience of seeing the original barn burn down the year he purchased the farm, and facing the daunting task of building another one.

Morrison must have been a visionary. The uniqueness of the barn and how it served his needs were ahead of his time. This large, 7-bent, 56-foot by 100-foot [17 m x 30 m] barn had several new features. There were two threshing floors accessible from the west side and a granary located in the centre of the main barn floor. The horses could bring in the crops, have them unloaded and then continue around the granary and out the other threshing doors. Morrison thought out such details as adding vertical 5-inch by 4-foot [13 cm x 1.2 m] dowels of hard maple around the corners of the granary. As the wagons turned the tight corner of the granary, they would roll off the dowels and damage neither themselves nor the structure.

Morrison also added a few other innovations in his barn. He installed built-in scales into the threshing floor, so he could weigh what he produced, a normal farming practice today. There were air vents connected throughout the beautifully built high stone foundation and two at each end of the roof. We know today that good air circulation keeps animals healthy.

Morrison built a corn crib that also served as part of the overhang in the forebay. (See photo 6) The forebay itself was unique, as only the central part was open to the elements, while either side was enclosed and led to the stables. Even the roof had an unusual design, with no pointed gables. (See photo 5)

The timber-frame bent design is a work of art. With the granary as a separate structure in the centre of the barn, the timber frame stands symmetrical around it. Posts, extra-long braces and girts seem to fly up in perfect unison to the 44-foot [13 m] peak of the roof. The beams are a typical Queen's Bush combination of maple, white ash, and pine, with rock elm serving as the sills.

Where the loose hay and straw were stored a wire mesh runs up to the top of the bent girt, about 24 feet [7 m] high, to keep the feed from falling onto the threshing floors. Two hayforks were installed, one in the centre and one off to the west side; both could travel the length of the barn and drop their loads behind the wire mesh.

By 1914 the Morrisons decided to build a greenhouse and begin a nursery business on County Road 1. They also built a beautiful cottage from round stones on the shoreline and,

14-4 *Not much has changed in almost 100 years from this shot taken in the early 1900s. The many trees that surround the barn were imported from Scotland in the 1880s. (courtesy of County of Grey-Owen Sound Museum)*

14-5 *The seven-bent barn built in 1896 had a unique forebay that was open in the centre with stables on the sides. The corn crib provides added protection as an overhang.*

14-6

14-7

14-6 *This corn crib is slatted in the front to provide good air flow to dry the corn.*

14-7 *One of the two ramps leading into the main doors. Just inside are floor scales used to weigh the grain and hay brought in. They were installed in the early 1900s.*

in 1919 severed from the farm 10 acres [4 hectares] that included these structures, and sold off the rest. In 1920 the Turner family bought the 190-acre [77 hectare] farm and continue to farm it to the present day.

"We were always in beef cattle," said Mac Turner, who just recently handed over the farm to his son, Bill. "My father bought 100 steers in the twenties and we've never looked back," he added. Mac has memories of being a kid growing up during the "dirty thirties". He recounts the tale of his father having to sell two-year-old steers in 1929 for $30 each, just after the stock market crash. "It got worse after that," he says. But like most farmers of that time, the family always had plenty of food on the table and all worked hard to keep that food coming in.

By the '50s the economy was booming and the Turners bought their first tractor—a 1952 Farmall Super C—and a plow for $1,750. Mac bought the farm in 1955 from his father and began an off-farm job later, as most family farmers must do today, as a real estate salesman.

There are still cattle in the barn, 100 stockers, the land is in production, and the barn stands tall and well maintained.

14-8

14-9

14-8 *The timber frame structure of the barn, made mostly from maple and white ash, is 56 feet wide [17 m] with unusually long braces connecting the top plate to the middle girt.*

14-9 *Inside the barn is a collection of odds and ends, even a Model "T" Ford. The light enters from the right through the slatted corn crib.*

15. The Stone and Timber-Framed Barn

Standing inside this granite stone barn I am reminded of the great stone churches of the middle ages. It is inspiring and magnificent in its building artistry and achievement of the pioneering spirit. What compelled a man to build such a structure in the wilds of Bentinck Township in the 1860s, we might never know.

15-1 *The stone barn built in the 1860s with timber frame additions to the north.*

The 250-acre [102 hectare] farm is located just north of Durham, known in the 1850s as the town of Bentinck. The first road to the area, the Garafraxa Road (Highway 6), was built in the late 1840s and carried many settlers into the Queen's Bush from Fergus, the last outpost of colonization at that time. Seventy-five miles [120 km] to the north was Sydenham, known today as Owen Sound.

The five original 50-acre [20 hectare] lots that make up this farm were all taken out of the Crown in 1849 by five different pioneers. In 1884, William George Jackson received 200 acres [81 hectares] of this farm by bequest; in 1890 he bought the 50 acres [20 hectares] to the south for $4,250, giving the present borders of the farm.

A large 40-foot by 80-foot [12 m x 24 m] traditional timber-frame barn, where most of the hay and straw were stored, was added to the stone structure in the late nineteenth century. George Noble bought this farm in the 1920s and for 50 years George and then his son Roydon cropped the farm, a mixed livestock and grain operation until 1971, when Roydon sold the farm because of health problems. Roydon, who still lives next door, is proud of the fact the farm produced such good crops over the years.

One of the favourite stops for early settlers was the British Hotel in Durham, also an imposing stone structure, that locals say was built in 1854 by the same mason as the stone barn just one-half mile north.

15-2 *The walls of the barn were made of predominantly granite stone gathered from the surrounding fields. The small addition housed a tank for water used for barn and house needs.*

The 30-foot by 60-foot [9 m x 18 m] free standing stone barn, has many timber-frame features incorporated into it. The stable has posts, main girts, and sleepers, notched together to support the floor of the mow. On the main level, two beams, about 10 feet [3 m] above the floor, connect the opposite walls of the barn to give them rigidity and form the borders of the threshing floor. Both of these timbers, called swing beams, are widest in the middle so as not to need any support under their 30-foot [9 m] span. Not needing posts in the middle gave more room to the barn floor area. The swing beams widen from 12 inches [30 cm] at their ends to 22 inches [56 cm] in the middle. One also supports the floor above the granary. We can tell that both these large timbers were set into the stone wall as the barn was being built since they extend into the walls. (See Photo 4) The granary is still in good shape and has six divisions in it to hold the different types of grain. Posts at the ends further support the floor above the granary that held loose straw.

The roof structure is an amazing piece of architecture. Every 8 feet are sawn 3-inch by 8-inch [7.6 cm x 20 cm] rafters, beautifully joined together with mortises and tenons at their peaks

15-3 *A variety of beautiful stone masonry, from large boulders to small flat stones. Contrary to what first might come to mind, the small openings were not for defense against marauders or thieves, but for air ventilation.*

15-4

15-5

15-6

15-7

and coming down to rest squarely on top of the stone wall. These rafters are further supported by short pieces of lumber embedded into the stone wall. On top of these are furring strips, then boards and finally roof sheeting. (See photo 5) The fact that there is no regular timber-frame structure to hold up the roof, speaks to the ingenuity of the builder, who has instead opted for what we would call a more conventional roof support system, while incorporating some precise notching at the peak. A fine rafter end feature is the short, carved 3-inch by 10-inch [7.6 cm x 25 cm] separate pieces of timber nailed to the sawn rafters on the inside. These rafter ends protrude outside like little sailing ships every eight feet. (See photo 6)

The stone walls of the barn rise to a height of 34 feet [10 m] at the gable peak. The walls begin with a bottom thickness of 30 inches [76 cm] tapering to 22 inches [56 cm] at the top. Most walls are made from granite boulders and rocks, with small cut stones at the corners, around the main door and a small door at the back. (See photo 7)

There were many, many stones used to build this barn and Roydon gave me a clue as to where they were gathered. He asked me to look around the nearest fields and see if there were any stone fences visible. No, but the neighbours across the road certainly had some. At first I thought Roydon was telling me the stone mason had to go farther afield to find stones, but I soon realized the stone fences that should have been around the fields near the barn were the stone walls of the barn. Roydon has plowed that land for years and knows the different fields like the back of his hand. The land slopes down from Highway 6 toward the back of the farm. There is good land at the back half of the farm, but with many small-sized rocks, almost gravel. But, at the top, where all the buildings are situated, one finds large granites, weighing between 10 and 300 pounds [4.5 kg and 136 kg]—perfect for stone construction. Instead of cursing these stones, the pioneer farmer found them a blessing, and the great stone barn was built.

On every wall there are eight to ten slits about two feet [60 cm] high. Many have told me, "they must have been used to poke a rifle through and fight off Indians and would-be thieves." (See photo 9) More likely the farmer who built the barn was thinking of ventilation for his crops, especially the grain. A traditional wooden barn is clad with boards. The wind

15-8

15-4 *The massive swing beam measures 30 feet [9 m] long and 22 inches [56 cm] at its widest.*

15-5 *Sawn rafters come to rest on top of the stone wall and are further supported by short pieces of lumber embedded into the mortar.*

15-6 *Beautifully shaped rafter ends support the roof overhang.*

15-7 *Nice stone detail on top of the main doors, supported initially by a plank framework.*

15-8 *All the corners of the barn are stacked with flat pieces of interlocking stone.*

15-9 *These openings were used for ventilating the stone barn which would accumulate more moisture than a wooden building.*

whistles through at every opportunity, providing plenty of ventilation to make sure a farmer's crop stays dry. In a stone-walled barn, humidity could be a problem, yet large window openings would not only compromise the structure of the barn, but let in more of the weather. Solution—window slits that are narrow on the outside, in this case 12 inches [30 cm], and wider on the inside by 6 inches [15 cm] to let in a continuous current of air. This system not only kept out most of the rain and snow, but having the slits wider on the inside accelerated the flow of air coming into the barn.

One of the first water rams in the area was built and used on this farm. Where the present owner has dug huge ponds to the north of the barn, Roydon's father had installed a water ram where he found a spring bubbling out of the hillside. He used the power of the spring and ram to pump all his water needs to a large cement tank, dug beside the stone barn on the south side. From there water could be gravity fed to the livestock through the barn complex and to the house, about 200 feet [61 m] away.

The stone barn and its large timber-frame additions sit empty, except for the storage of some household goods, loose hay, straw, and dust. A fatality of modern farming practices, the structures just don't fit into the present cash crop operation. The current owner, Norman Wilmer, has every intention of fixing a large upward crack in the east gable wall of the barn. Now the width of a closed fist, this crack has enlarged every year. The threshing floors of the stone barn and its addition have lost their lustre, and cracks have appeared where rain and snow have come through the main doors.

This farm must have been a showpiece in its day, and although the barns are not being used, the land is still producing bountiful crops.

16. Orchard Barns

There are many instances where barns built by the same builder are next door to each other. Although the shape and type of wood used may be different, framing clues help us see the similarities. Such is the case of the two barns in the apple-growing country of the Beaver Valley near Clarksburg. (Photos 1 and 3)

Lot 31, Concession 11 in Collingwood Township has two 50-acre [20 hectare] farms with comparable barns. Both are narrow, 35 feet by 55 feet [11m x 17m] and have similar timber-frame characteristics. The four bents are in the same configuration with the granary located at the northeast corner of each barn. Located on the east sides of both barns are large, double doors opening out into space. Not a common fixture, the doors allowed straw to be blown into the barn during threshing or pitched in later with forks. The foundations themselves are also similar. At the top of each stone wall there is a layer of flat shale, acting as shims to determine the final height of the stables.

Photo 2 shows a common scene in 1905: horses, men, straw, and a barn. The straw is ready to go into the barn via the threshing doors and some fine horses are caught for posterity through the eye of the lens. The barn today is surrounded by trees and where the straw stack is in the photo, a pond has formed with a very large willow tree beside it. What have brought the greatest changes to the uses of the barns are a dairy operation and orchards.

The west half lot of these two farms was bought around 1915 by William Camplin. He ran a mixed farm operation, as was quite common in those days. Unfortunately, that same year his wife died while giving birth and thirteen years later William succumbed to tuberculosis and left his enterprise to his only son, 13-year-old Herbert. Herbert was raised by his aunt and uncle, who farmed across the road and also worked the Camplin farm. In 1936, when he was 21, Herbert and his new bride moved in and began a dairy and apple business.

16-1 *This barn and the next door neighbour's barn (the Camplins') were built by the same framer. Each received an L addition which was later taken down. Behind this barn is a modern facility housing a dairy operation.*

16-2

16-3

The Camplins' farm is in the heart of the Beaver Valley apple-growing area, which gets its mild growing seasons from the warming effects of Georgian Bay and protection from severe cold by the valley itself. For some time apples from this famous valley supplied the Queen of England's table.

In 1938, Herbert planted 8 acres [3 hectares] in apples—Macintosh, Spy, Wealthy, and St. Lawrence. Herbert's son, Ross, remembers travelling with his father when he was 16 years old, to towns like Shelbourne and Midland, peddling apples door to door from the back of the family's truck. The apples would be taken out of storage in the cold cellar of the barn, packed in six quart baskets, and put in the truck for delivery. One small wholesaler was supplied with Camplin apples for more than 30 years.

"There were good prices for good quality apples," said Ross, "and you know, we would truck all winter in all kinds of crazy weather." After Herbert got older, Ross took over the driving, his father riding right next to him, selling apples to long-time customers until Herbert passed away in the mid-1990s. Four acres [1.6 hectares] of the original orchard remain, and Ross continues to pick from those same 60-year-old apple trees, which still bear good fruit.

Herbert tried a little bit of everything. He grew one-half acre of raspberries, hired people to pick

16-2 *The barn in 1905 after a few years in use. The large straw stack in the foreground would either be pitched into the barn through the double doors seen just behind the stack or else left to winter outside in its conical shape designed to keep off most of the weather. The barn board looks fresh and the roof is straight. (courtesy of County of Grey-Owen Sound Museum)*

16-3 *More than 100 years later, the barn stands intact, although a little worn.*

them and sold them by the crate [24 boxes to one crate) to local grocery stores. He also had success with commercial tomatoes and chickens and supplied local inns with both.

He was one of the first in the area to grow corn for silage. In 1948 a crew built a 12-foot by 38-foot [3.6m x 11.6m] concrete silo. "It was the biggest around," remembers Ross, "and then to prove he needed such a large silo he had to always grow enough corn to fill it." One of Ross's first jobs was to tramp the corn down as it came into the silo from the self loader.

The corn was needed for the heart of the Camplin farm—the dairy. From a modest start with 2 dairy cows in 1936, to hand milking 13 cows in 1944 to a milking parlour and bulk tank in 1966, dairy continued to be the mainstay of the farm until Ross broke

16-4 *A hay fork used in the Camplin barn. The points would be facing down and driven into the loose hay and then tripped by a rope and the hay would be lifted. This photo also shows where the foundation was continued for the addition.*

16-5 *The original framer's markings [18 2] still show on the Matthew barn frame. Each timber was marked by the master framer so that the right piece was fitted into its corresponding timber.*

16-5

16-4

16-6

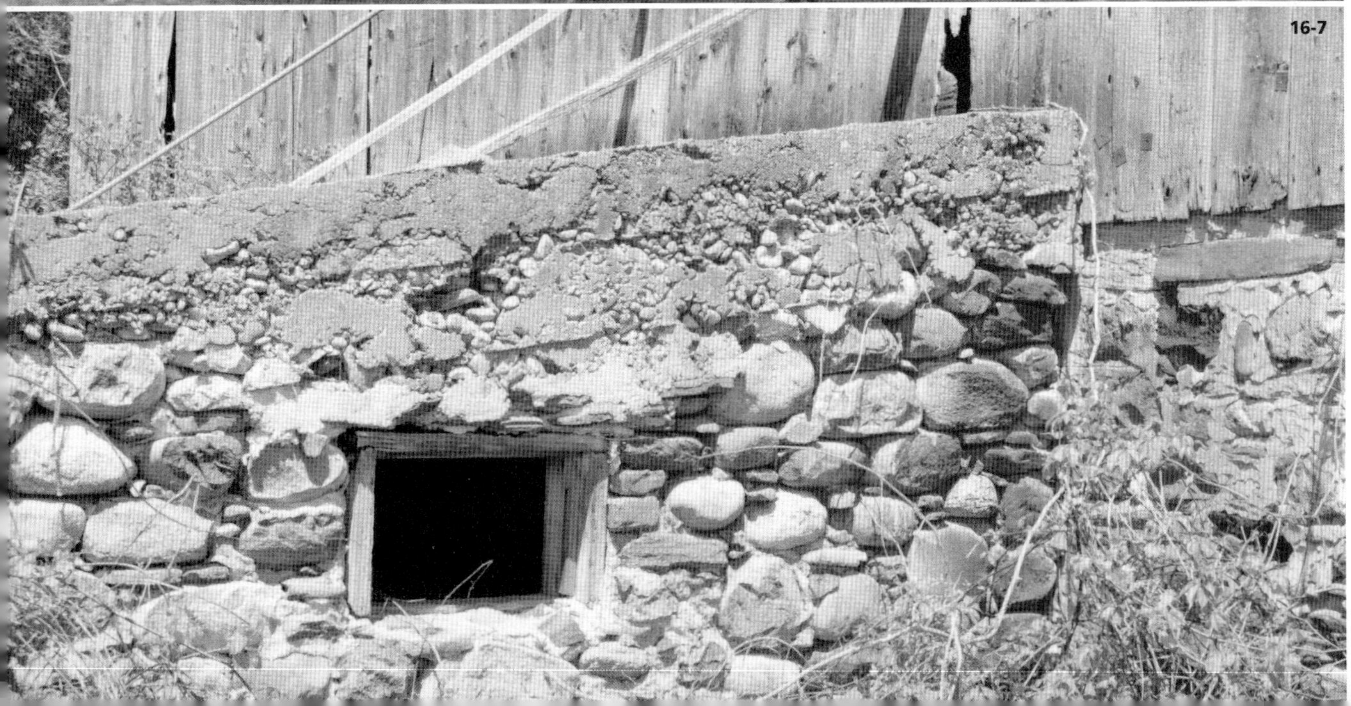

16-7

his back in a farming accident in 1996 and sold the quota and livestock the next year.

During the 1940s milk would be shipped every morning in 8-gallon [36-litre] cans to a dairy in Thornbury. Any excess milk would be separated and the cream sold to a local creamery as well as to creameries in Tara and Owen Sound. The Tara plant was bought out by Gay Lea in 1966 and they began picking up milk in the Beaver Valley area.

"The two biggest changes were the Surge milking machine we bought in 1944, when we got hydro and the milking parlour we installed 20 years later," said Ross. When milk quota came up for sale in 1968, Ross bought some for $5 per hundredweight.

On the east half of Lot 31, Concession 11, the 50-acre [20 hectare] farm was originally farmed by the Matthew family who had the barn built in the 1870s (Photo 6) and addition put on in 1908. The Watts family took over the farm in 1936, when part of it was planted into an apple orchard. A root cellar under the gangway of the barn stored the apples and some vegetables. (Photo 7) When Don Watts acquired the farm in 1960, more and more land

16-6 It all began with a barn—*The Matthews' barn, raised in the late nineteenth century. After the initial barn raising, the owner usually finished off the siding and roof with the help of a few neighbours. This is a classic four-post, four-bent timber frame structure. Horses in the foreground were the work beasts of that time and helped as much with construction as did humans. (courtesy of County of Grey-Owen Sound Museum)*

16-7 *Building a root cellar under the gangway leading up to the threshing floor was a common practice, seen in many barns of the Queen's Bush, and here at the Matthews' barn.*

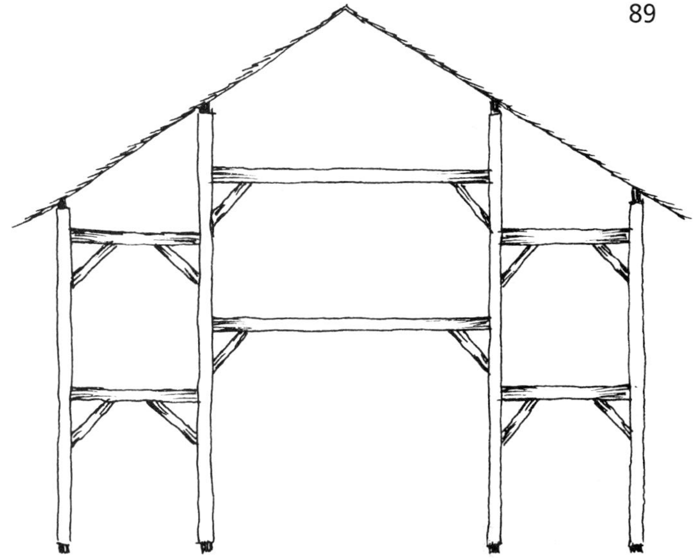

was planted into apple trees, until today most of the 50 acres [20 hectares] are orchard. For a while, much of the crop was trucked to a juice plant in Thornbury.

Their neighbour, Ken Almond, remembers Don as a "city boy who married a country girl but was a good, hard working man, who was always interested in learning something." Although Don held a professional job in Owen Sound, he managed to accomplish much on the farm. "He used to come over a lot and I'd go over and show him how to plow or cultivate, and advise him when to do it. But back then that's what we did for our neighbours—we helped each other out, and it sure wasn't for the money," said Ken.

The addition to the barn built in 1908 was taken down years ago (See photo 8). The barn addition marked another milestone in the Matthews' enterprise, accomplished with the help of the whole community.

The raising day must have been an important social event to get so many women in their finest dress—women who probably had farm chores, children to care for and food to prepare back home—to come for a picture of a barn raising. What were their thoughts, as the picture was snapped?

16-8 *A large L addition is being added to the main barn in this 1908 photograph. The women are all dressed in their finest for this dramatic shot. (courtesy of County of Grey-Owen Sound Museum)*

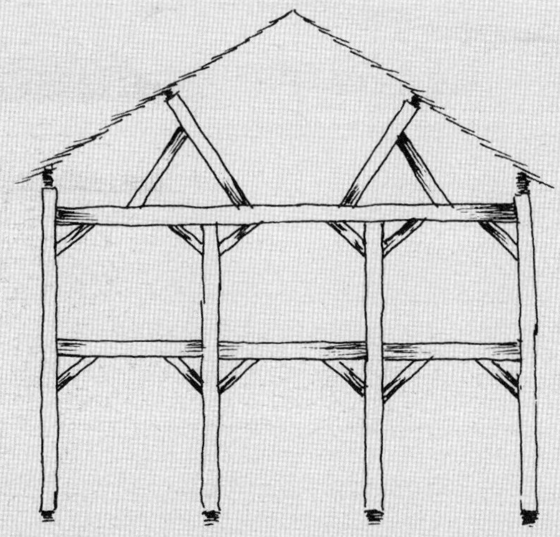

17. Twin Barns

At the southern edge of the Queen's bush, near Lucknow, one finds rolling land dotted with houses and barn foundations built by Scottish stone masons. The town was named after Lucknow, India, where in 1857, a 4,000-strong British army went to the relief of Europeans besieged in that memorable city in India. The native population in and around Lucknow rose up against the hold Britain had at that time on the whole country. The relief army, made up mainly of Scots, managed to rescue some 1,600 Europeans and defeat a 40,000 strong Indian army. It was big news at that time and Balaclava, as it was first known, became Lucknow.

The first Scottish pioneers to the area were seasoned masons and built some very fine stone buildings and barn foundations.

Just west of town, along Huron County Road 86, which is also the division between Bruce and Huron Counties, lies the Larry Henderson farm on the south, or Huron County side of the road. The land is characterized by rolling hills with good patches of productive land. There are two nineteenth-century barns on this site, unique in the way they overlap each other by about 6 feet [1.8m], just enough for an entrance between the two.

Both barns are almost in original shape, as the previous owners, two brothers, James and Charles Baird, had lived out their lives on the farm without changing very much. In fact, until Larry bought the farm in 1998, there was just one light fixture in each barn, and three in the old stone house.

The back barn, measuring 36 feet by 56 feet [11m x 17m] and the front, which is a little bit larger measuring 40 feet by 60 feet [12m x 18m] were built in 1869. Because of some uncharacteristic use of beams, I believe both timber frames were notched by the same framer.

Inside both barns the top girts on either side of the threshing floor have another beam notched into the posts about three feet below them. These beams certainly gave extra strength to the structure, but would require more notching, more material and weight in raising in each bent. There is another girt about knee level, spanning the width of each barn. Larry proposed the idea that in the days of loose hay, boards could be placed between the upper and the lower of two girts to keep the hay or straw from spilling out onto the threshing floor. (See photos 3 and 4)

Because of their unchanged interiors one just has to look up inside the barns and see near the very top of the roofs, attached to the ridge timbers, the small hardwood beams which run the length of each barn and still hold the hayforks.

The barns' post and beam structures are made from local wood—8-inch by 10-inch [20 cm x 25 cm] elm for the roof girt supports, 12-inch by 12-inch [30 cm x 30 cm] pine posts for the four centre supports around the threshing floor, and assorted woods such as maple and tamarack for the bent girts, joists, and girders.

What is unusual, of course, is the narrow attachment of the barns, only six feet [1.8m]. At one time there was a large addition in front of the second barn and attached to the first that filled up the space of the common loafing yard, taking advantage of a protected space between the two barns.

Originally there were two ramps, both on the west side leading up to the threshing floors. Neither barn is banked into a hill, so the ramps were fairly steep and hard to maintain. The back barn's ramp was taken out when a new silo was built and in 1999 the first barn's ramp was also taken out, because tractors had a hard time backing a loaded wagon up the steep gangway. All hay is now put into the barns by way of elevators, a method that suits Larry well, as he said the floors of the barns were not meant to carry the weight of today's tractors.

Another unusual feature is that the first barn had only one-half of a stone foundation; the rest of the frame sat on mud sills with regular

17-1 *A view from the east side shows two large timber-frame barns on the Henderson farm built by the same framer at the same time. Why were they built kitty-corner to each other? Perhaps we'll never know.*

17-2 *The barns overlap by only 6 feet [2 metres] with a small entranceway between the two mows and one between the stables.*

sleepers and a floor on top. Larry excavated the rest of the stables and poured a concrete foundation under the other half, doubling the size of his stables. When the barns were first built, the half-stable of the first barn was used for cattle and the second barn housed horses and perhaps pigs.

Two barns kitty-corner to each other with only a small attachment... I would have guessed that one was moved to that site at some past date, but that was not the case, as Larry assured me they were both built at the same time. Why the original farmer built in that way, perhaps we'll never know, but today they are two more examples of excellent timber-frame construction, side by side.

17-3 & 17-4 *One of the clues that it was the same framer who built the two barns: two girts, instead of the common one, span the width of both barns.*

17-5 *Looking at the barns from the west, this photograph shows both ramps up to the threshing floor have been removed. One can still see the large main doors of the barn on the left but the doors on the other barn have been boarded up.*

17-6 *This south-facing stone foundation is showing its age after 130 years.*

18. The Leaning Barn of Crawford

It's hard to see this barn from the road because of a row of trees. But through a space between two trunks, if you are not travelling too fast, you will notice the barn is leaning slightly to the north. This well-built timber-frame barn, constructed in 1904, was hit by a tornado 61 years later in 1965 that destroyed two neighbouring barns, and left this one with a slight lean.

The Bentinck Township farm was originally taken out of the Crown by Archibald McDougall in 1892. It stayed in the McDougall family until 1998, when it was sold to the present owner. Ina (McDougall) Cordick, whose father farmed the land from 1938 until 1988, has happy memories of growing up on the farm.

"It was a lot of hard work, but it was satisfying. You felt like you had many triumphs," she said. The hard work and triumphs mainly centred around the barn and bringing in the crops.

18-2 *Almost 100 years after the raising the people are gone but there are still echoes in the barn of that joyous event.*

Behind the McDougall barn the land rises gently but steadily, making haying a risky job. Ina Cordick remembers sitting on top of the hay wagon, which was being pulled by horses, looking out for groundhog holes. Because of the steep slope of the land, if a wagon's wheel dropped into a hole it would be just enough to tip over a whole wagon load of hay and its occupants.

18-1 *OPPOSITE PAGE. The McDougall barn was raised in 1904 by a large contingent of neighbourly help. The smaller attached barn frame was built at the same time as the main four-bent frame. (courtesy of Ina Cordick)*

18-3

18-4

Ina recalls many chores: milk the cows, feed the pigs, the horses, the chickens and the occasional duck. The cream separator was hand cranked to provide cream for churning the household's butter.

"One of the hardest jobs my father had was forking the loose hay out of the mow. It was packed so tight it was really hard to get the fork in there and then pitch the hay down to the threshing floor. From there the hay had to be stuffed through the hole in the floor to the animals below."

Over and over Ina emphasized that farming was a good way of life. "I guess life isn't just about money," she said. Ina had three siblings. "We would play under the outside barn light almost every night until my mother would call us in. We just made up a lot of games and it was so much fun. We had such happy childhoods."

It wasn't until 1947 that electricity was connected to the farm and not until the 1960s that pressurized water was installed. Until then the windmill pumped water, gravity-fed to the barn and house. Cisterns were installed at the house and the barn. "That was one of the chores, getting water.

18-3 *The lean of the barn is to the north after a devastating tornado in 1965 wiped out two other barns across the road. Unfortunately, the family apple orchard behind this barn was totally uprooted by the tornado and never replaced.*

18-4 *The land rises steadily to the east of the barn making haying a sometimes precarious job. As a young girl, Ina Cordick would sit on top of the hay wagon and look for groundhog holes, which could spell disaster for a fully loaded wagon.*

18-5 *Beautiful example of a queen-post timber frame roof support. The ladder leads up to the hayfork and track to provide access for repairs.*

18-6 *One of the reasons the barn withstood the tornado in 1965 were the double braces located at the ends.*

18-7 *A well-made foundation with large blocks for the corners. The end sill is plainly visible on top of the foundation.*

When I was a little girl I almost fell into the house cistern. It's funny now when you remember."

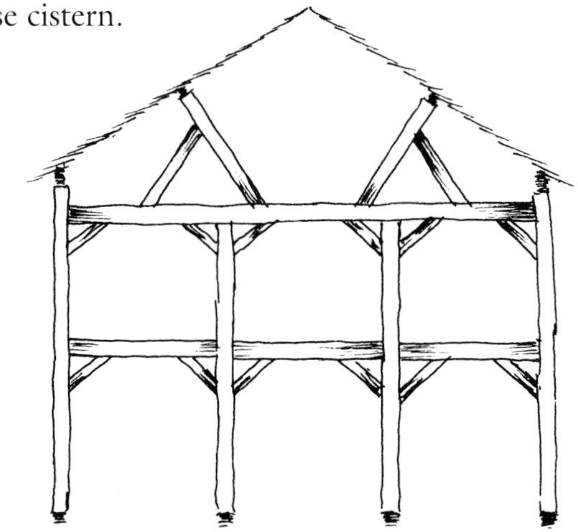

"The neighbours used to get together a lot more back then. They always helped each other out," Ina said. There were work bees for anything from threshing to quilting. After the destructive tornado of 1965, all the neighbours gathered in a building bee to help re-construct the two destroyed barns.

With the introduction of tractors farmers became more independent, but "doing hay" still meant any student could get a job in the summer on the back of a wagon loading square bales right up to the 1980s and '90s.

This barn is in the shape of an *L;* the larger barn frame measures 40 feet by 55 feet [12 m x 17 m] and the smaller 30 feet by 50 feet [9 m x 15 m]. Unfortunately, where the two barns meet, the roof has leaked for many years and the water has already rotted a couple of elm sleepers which have fallen into the stalls below, making the floor unsafe. The stalls stand empty of livestock, filled in with cobwebs, old manure, and bits of wood and junk.

Broadaxe marks are evident on the beams and the top four connecting bent girts are 40 feet [12 m] in length, 10 inch x 12 inch [25 cm x 30 cm] maple. What great trees they must have been and all cut from the farm bush. One of the reasons this barn withstood the tornado of '65 was that most of the posts and girts were made from hard maple and there were double braces at the ends of each bent. The old hayfork still hangs in the larger barn, waiting for the day it will be swung down when the horses pull in another load of loose hay.

18-8 *The autotruck was an early steel-wheeled car used as a tractor for pullng small wagon loads. Production of farm implements was interrupted during the Second World War; machinery was not available again until 1946, and so people made do with what they had. (courtesy of Ina Cordick)*

19. A Different Barn Roof-Line

When Elmer and Marguerite Golem took over this Sullivan Township 100-acre [40 hectare] farm in 1942, they could only afford to rent it. The previous owner, Alex White, had lost the farm during the Great Depression of the 1930s, when he couldn't keep up with the mortgage payments. The farm reverted to the original owners, descendants of the Kyles, who had bought the farm in 1864.

In 1945 the Golems found themselves in a quandary. The farm was up for sale but they could not afford the $4,500 asking price. Elmer was working as a hired hand on another farm and Marguerite had quit her $550-a year teaching job, as married women were not allowed to teach. Banks did not offer mortgages in those days, so Elmer, a hardworking man with a good reputation, took to the road, asking neighbours if they could lend him the money. There was no extra money in farmers' pockets in 1945. People were still reeling from the Depression and the scarcities of World War II. One day Marguerite and Elmer stopped at the gas station in Scone owned by the Sulkyes, and Elmer asked the owner if he knew anyone who might lend him some money. The owner thought a minute and said, "Well, I could do it." "It was providential," said Marguerite.

19-1 *A typical 1940s picture of farm life shows Percheron horses (one called Nellie) pulling a stone boat with the Golems' two sons and a load of potatoes. Note the brick pump house building in the background and new power line installed on Christmas Eve in 1947, a few months before this picture was taken. (courtesy of Marguerite Golem)*

19-2 *Here is Elmer's alternative to buying an expensive square baler in 1950—a buck-rake attached to his first tractor. A good view of the inner courtyard of the three barns which hasn't changed since then. (courtesy of Marguerite Golem)*

The Golems lived on that farm for the next 45 years, until Elmer passed away in 1987. "It doesn't seem long working with a partner who is so compatible," Marguerite said.

The original barn, the one leaning in picture 4, was built by Joseph Kyle in the early 1870s as a "practice barn." "They were teaching themselves how to build a barn, so that one didn't turn out so well," Marguerite remembered from the story passed down through the years.

Looking at the inside of the 30-foot by 40-foot [9 m x 12 m] mainly hardwood structure, I judged the principal cause of the lean is missing knee braces. Many of the braces on the east—the side to which the barn leans—are just not there. Precision plays a very important part in timber framing. Final measurements are calculated in sixteenths and the notches for these braces are off by one-half to almost one inch for the 36-inch [91 cm] braces. (See Illustration 2.)

The "practice barn" did pay off, as the main barn, measuring 55 feet by 60 feet [17 m x 18 m] was built from hardwood beams and has stood the test of time. In fact, on the farm north of the Golems', a similar barn was built by Joseph Kyle's brother John. Probably

19-3 *The main barn was built in 1890. All the trees around the house on the left were planted by Marguerite Golem during the 1960s.*

19-4 *This barn has many unusual roof angles due mainly to the silo's accommodation inside the barn. On the extreme left is the first 'practice barn' built in the 1870s, which is still standing with quite a distinct lean as can be seen even 50 years earlier in photo 2.*

it was a co-operative effort, with the brothers building the first barn in 1890 and the other shortly after.

The main barn was built on a gravel hill, with the gravel first used as a mix for concrete for the imposing stone house, the driveway, and then the stone foundation for the barn. Once enough gravel was scooped out to make room for a stable, the main barn was constructed and raised with an attached driving shed and stable underneath. This driving shed would eventually house a concrete silo built by Elmer in 1954.

Co-operation and sharing the work with neighbours was a big asset for the Golems. During the 1940s Elmer, with seven neighbouring farmers, bought a thresher to be shared by all. Records were kept of how long each used it, while all eight farmers helped one another gather in the grain. When square balers came in the early 1950s, the neighbourhood got together on that purchase also and the County Line Forage Company was born.

But standing on the back of a wagon stacking bales for himself and the neighbourhood proved to be too much work and time for Elmer. He had to quit and look for another way to harvest hay without the great expense of buying a baler. First he adapted a buck-rake (See photo 2) on the front of a semi-tractor he purchased, called an autotruck. With a buck-rake he could pick up the windrows of hay, "quite a load" as Marguerite said, take it into the barn, and drop the load onto slings. Once there was a full load on the barn floor, he would hoist the hay by using the hayfork attachment and the slings to drop it on either side of the threshing floor. This was still a savings in time compared to forking the loose hay onto a wagon. The barns were tall, and accommodated the loose hay very well.

He then bought a McKee Harvestore, a forerunner of the forage wagon, except dry hay was blown inside the wagon, compressed and unloaded into the barn. Finally, by the late 1960s, when square balers became more available and less expensive, Elmer bought one.

"We feared the coming of big industries, controlling products from beginning to end," laments Marguerite, commenting on the state of farming today. The 1950s CBC "Farm Forum" radio show was an inspiration to the Golems and their neighbours. Farmers would gather for the half-hour show and a discussion would ensue about the topic of the day. Out of those discussions came the fierce independent thinking of sustaining family farms and receiving a fair share of the financial pie from farm products. Grey and Bruce Counties had more farm forums than any other region in Ontario, with more than 200 groups participating. The adult radio education program began in 1939 and ended by the late 1960s, when interest and activity dwindled.

There were originally two windmills on the farm, although both had been torn down before the Golems arrived. The first was located in front of the house and pumped water to the little brick building (See photo 1) for storage. From there the water flowed by gravity to

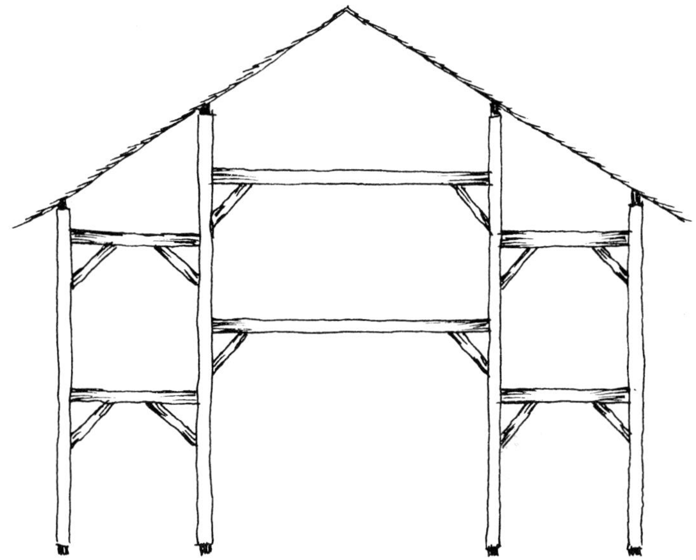

Illustration 1

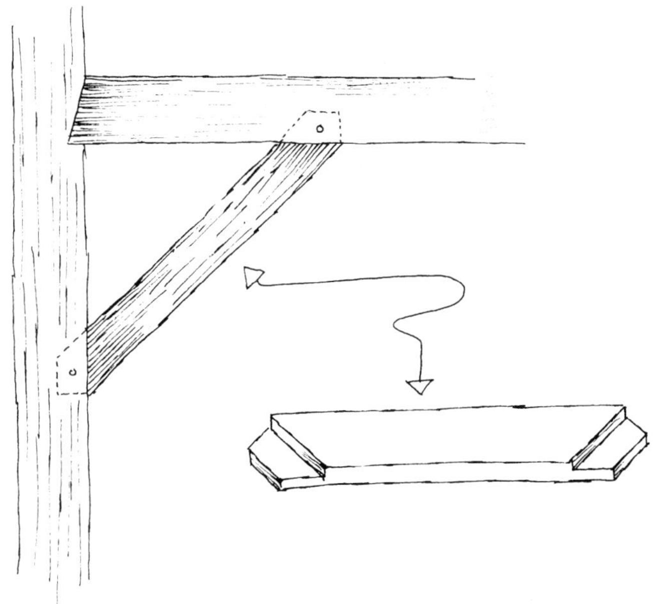

Illustration 2

19-5 *The silo is housed inside what was originally a driving shed with a dormer built to accommodate the height of the structure.*

the barn. Water was carried into the house in pails until electricity came to the farm in 1947. The other windmill was situated on the roof of the main barn and was used to grind grain and corn. A storage cellar for turnips, built in the stables of the driving shed, was demolished by Elmer to make way for a tractor door into the stables on the west side.

On the main barn, an odd dormer sticks out of the east roof about halfway down. Underneath it and inside the barn is a 25-foot [7.6 m] concrete silo. Elmer hired Louie Kaufman, who had forms for silo construction, and with the help of neighbours, built the 10-foot [3 m] diameter silo in one week. The concrete was mixed by hand and hauled up on the inside of the silo with a rope and pulley attached to one of the beams of the barn. "Constructing it inside the barn was a good way to keep the weather out of it. It made good

19-6 *Inside, on the second floor of the driving shed we see the top of the 10-foot by 25-foot [3 m x 7.5 m] concrete silo built in 1954. The original roof was taken off and a new dormer built. A fine example of a queen post timber-frame bent.*

corn silage," said Marguerite. In photo 2 the large open door closest to Elmer shows the location of the silo built four years after this picture was taken.

"We made our living from the 100 acres since 1947 and raised three children as well," said Marguerite. "You can't say that about the state of farming today," she added. The Golems began as a traditional mixed farm—chickens, pigs, some beef cows and a couple of milking cows. Eventually, the dairy was expanded and, by 1951, when Elmer installed a pipeline for milking, he had built enough stabling for 17 dairy cows.

Milking cows became a "good way of life," Marguerite said, "and what made it possible was the help and co-operation of neighbours. Helping your neighbours was like having time in the bank."

19-7 *The back of the main barn and the first 'practice barn' on the right.*

19-8 *Foot holes in the side of the granary lead to the first rung of a ladder above.*

20. The Symphony Hall Barn

While dairy cows low in the stable below, bells jingling around their necks, a symphony orchestra might be playing in the mow, or an opera star singing, a rock and roll band blaring or famous chefs from Toronto serving up culinary delights. It all adds up to a sometimes colourful figure, Michael Schmidt, presenting his version of the versatility of a barn.

In 1989 Michael wanted to expand his dairy operation and planned a large addition to his barn, originally built in 1866. He decided on the time-proven method, unusual these days, of a timber-frame addition. I was the framer and, with my friend Peter Rhebergen, took down another barn that was not being used and began to notch a 40-foot by 60-foot [12 m x 18 m] frame, using the old—but sound—timbers.

Michael joked that he didn't want any centre posts so that he could create a space for an audience at future concerts in the barn. I really didn't believe him, but the customer is always right, so after a traditional timber-frame stable and floor were built, strong enough for a tractor to run on, we decided to incorporate some new technology in the bents. I notched two, 40-foot [12 m] bents, one for each side, and one long 60-foot [18 m] bent for the end wall. There was no bent in the middle to hold it together. What we used instead were roof trusses, fastened from the original barn roof and timber frame to the 60 foot [18 m] end wall, to give the new timber-frame support and strength.

There was an old-fashioned barn raising on the big day. Friends and neighbours came, but without the traditional tools and knowledge and with such long and tall bents, we used a crane to help us. A crew of Amish men from Sullivan Township completed the roof, siding and walkway above the stable. (See photo 3)

Michael realized his dream of concerts and organic gourmet dinners, and I fulfilled my dream of notching a large traditional barn. Since then, Michael has hosted the Symphony in the Barn series, with the full Kitchener/Waterloo orchestra playing such classics as Mozart's "Magic Flute" and Haydn's "Creation".

20-1 *Michael Schmidt, conductor and farmer. A timber-frame addition in the early 1990s provided a "symphony hall" for Michael and a place to store hay during the winter. Straight above Michael's head is one of the two curved beams made especially to accommodate the hay fork. (courtesy of Pief L. Weyman)*

20-2

20-3

20-2 *A photo in the 1970s before the addition. The smaller silo was taken down. (courtesy of Tom Firth)*

20-3 *The 40-foot by 60-foot [12 m x 18 m] timber-frame addition with a cupola on top is attached to the main barn, originally built in 1866. The seating for the shows held during the summer months is in the addition while the main stage is in the old barn.*

20-4

20-5

20-4 *The main barn door features a painting of the sun. The lettering* Roots in the Rafters, *refers to a CBC radio show recorded there in 1998.*

20-5 *Inside the original barn and to the side of the stage, a second floor above the granary is used as part of the seating. Note the swing beam above the carriage is 36 feet [11 m] in length and 18 inches [46 cm] wide in the middle.*

20-6 *When the Firth house burned down in 1890, the Firths moved to the Grange Hall across the road. The building was taken down in the 1960s. (courtesy of Tom Firth)*

In 1998 Philly Markowitz, host of the CBC radio program "Roots and Wings", recorded an eclectic night of African and Arabic music to a full audience who danced until the dust rose thickly to the rafters. The barn has been the setting for fund-raisers for a local school, a theatre and many square dances.

In the fall, the stage and seating areas revert to more traditional uses and are filled with round and square bales of hay and straw.

Innovative ideas seem attracted to this barn, which has up to now gone through three major reconstructions and additions.

In 1848, Joseph Firth and a friend came up to Glenelg to claim two 50-acre [20 hectare] lots. It turned out his friend had a lot that was half swampy and not conducive to good farming. Joseph talked his friend into giving him the 50 acres [20 hectares] for a yoke of oxen, resulting the 100-acre [40 hectare] farm we see today, located 7 kilometres northeast of Durham.

Joseph must have been an industrious man because by 1861 township records show 75 acres of cleared land, with an assessment of $2,500. The tax assessor's list records the amount of produce and its value on the Firth farm for 1861:

25 acres fall wheat	500 bushels
7 acres fall oats	200 bushels
1 acre potatoes	200 bushels
7 acres fall turnips	2,000 bushels
10 acres hay	10 tons
wool	30 pounds
maple sugar	100 pounds
butter	300 pounds
bulls or oxen	2
heifers or steers	14
milk cows	4
horses	2
pigs	7

TOTAL VALUE OF THE LIVESTOCK — $515.

20-7 *In 1958 Tom Firth decided to sheet the roof with aluminum. It took days to put on the heavy-gauge sheets but more than 40 years later, the roofing is still in good condition. (courtesy of Tom Firth)*

20-8 *Tom Firth on a sled in 1928. The barn is unpainted and has only one silo and part of a chicken house showing. (courtesy of Tom Firth)*

By 1864 the Firths moved out of their first log cabin and into a frame house. Two years later, the first 36-foot by 60-foot [11 m x 18 m] timber-frame barn was built, resting on mud sills. There must have been other buildings to house the livestock, since the barn was used mainly to store the winter supply of hay, straw, and grain. Twenty years later, in order to house the livestock, the structure was raised, and a stone foundation with stabling built under it. A common practice in those days, raising the barn speaks wonderfully of timber frames, their great strength and flexibility as buildings, to be able to be lifted, adjusted, and moved about. Remember, there were no cranes or bulldozers or powerful machinery to help in these endeavours, just people's ingenuity, manpower, horses, and the effective, simple tools of levers, pulleys and winches.

Hard luck struck in 1890, when the Firths' house was destroyed by fire. The family moved across the road to the Grange Hall until the present brick home was built. Grange Halls were a precursor of today's farm supply co-ops, and some argue the Grange was a true farm co-op, where farmers got together and sold their produce in bulk for the best price. They also purchased goods in bulk, such as farm equipment, everything being shipped by trains. The nearest railway station to the Firth farm was in Durham.

One of the greatest innovations in the Firth barn came about in 1916. By then the new owner, William G. Firth, Joseph's nephew, wanted to expand the operation and for that he needed a larger barn. In fact, he tripled the size of the original structure.

Loose hay and straw took up a lot of room and one of the easiest methods of creating more storage was to add a new hip roof. William did more than that; he also added another 18 feet [5.5 metres] in width, making the overall dimensions of the barn 54 feet by 60 feet [16.5 m x 18 m]. The whole roof was taken off, including the purlins and plates, and a new roof structure, with the hip, was built.

The evidence of the expansion is very visible inside the barn. The new purlins and addition timbers have saw blade marks on the wood while the older frame bears broadaxe marks.

20-9 *The addition features a second-storey, outside, covered passage-way leading to the main entrance for concerts and other events.*

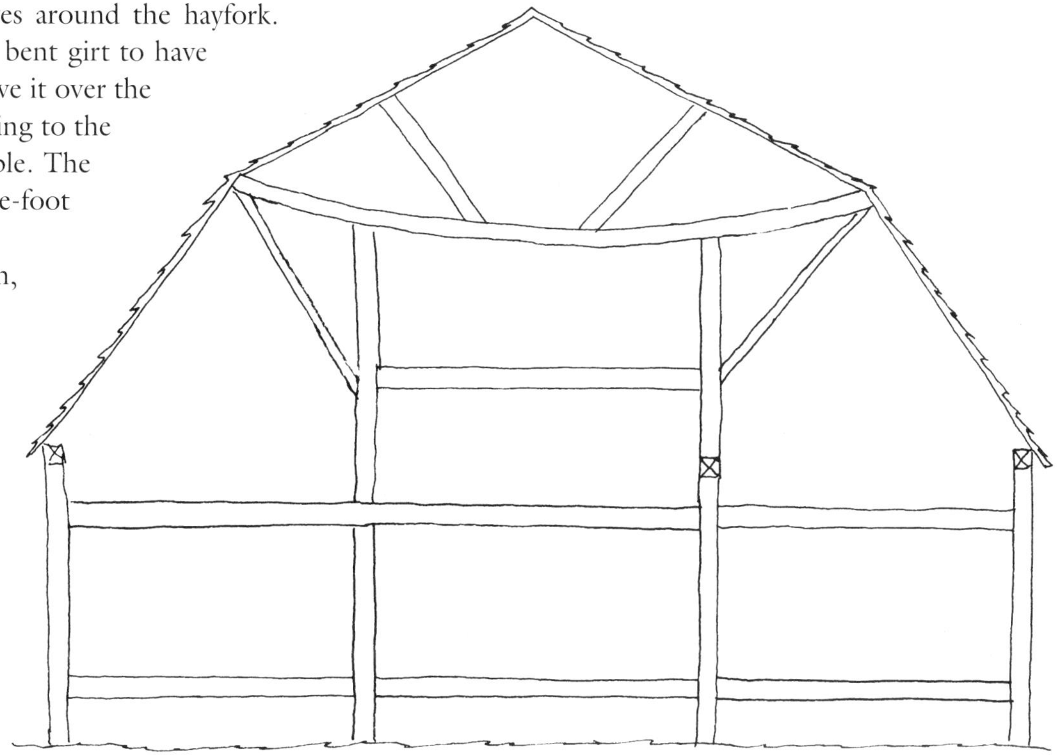

One of the most remarkable stories about this frame revolves around the hayfork. There needed to be a total of 15 feet [4.6 m] down to the first bent girt to have enough room to attach the hay to the hayfork and be able to move it over the girt. The top girts were part of the two middle bents and, according to the plan of the barn, there was only 14 feet [4.3 m] of space available. The solution was to find two trees, 18 feet [5.5 m] long with a one-foot [30 cm] bow in the middle.

"It took a long time for them to find the right trees," said Tom, the last Firth to own the farm. "It was winter, and the men trudged through the deep snow looking at all the trees in the back of the farm." Eventually, the right trees were found—two maples—and specially cut into beams with an axe and an adze, to keep a one-foot bow in the 18-foot [5.5 m] span. The girts can be still seen up in the barn, two bowed beams, with enough clearance for the hayfork to transport the hay around the barn. Now that's ingenuity!

A nervous bunch of men gathered on the day of the barn raising. The new sections of beams had to be raised on top of the old barn frame, which was already 18 feet [5.5 m] high. Some said the new bents would be blown down by the wind, others that it was not strong enough. The bents were raised, and stand solidly today.

There were still trials and tribulations that the whole community faced when the weather turned nasty and when the economy turned for the worse. Tom Firth remembers as a young boy having only potatoes and turnips to eat one winter instead of bread, because the severe fall weather froze the wheat crop and there was no flour, or money for bread.

Tom took over the farm in 1956 and was one of the area's farmers instrumental in bringing the system of milk quotas to Ontario dairy producers.

In 1965 Tom and three others complained to the Combines Commission that milk processors were forcing the Durham-area producers to ship only to them, creating a monopoly, which was illegal under the federal *Combines Act*. Hearings were to take place in Stratford but the processors backed out; dairy farmers speedily formed the Ontario Milk Marketing Board.

By 1966, Tom had up to 40 milking cows with 36 tie stalls. He began his own Holstein breeding program, picking and choosing the best cows. He continued milking until he sold the farm to orchestra conductor and farmer Michael Schmidt in 1983.

20-10 *An ornate cupola stands on top of the addition built in the early 1990s.*

Michael turned the farm into a biodynamic operation, milking cows and growing organic hay and grain. He built a bakery on the farm and presently sells fresh bread and other delectables on Fridays and drives to Toronto with his baked goods two other days of the week. It's the niche market that for some farmers provides a way to survive.

At the beginning of this century once again the sounds of music from an orchestra could be heard wafting from the barn, with cowbells jingling along to the tune.

21. Timber-Frame Silo and Barn

On Bruce Road 1, three concessions north of Lucknow in Kinloss Township, a picturesque farmstead with an unusual-looking barn catches your eye. This well-kept barn has a narrow, but very tall, addition on the west side—a late-nineteenth century timber-frame version of a silo.

Although square on the outside, the 28-foot [8.5 m] high structure has round metal hoops on the inside. To fill up the 14-foot by 14-foot [4m x 4m] silo, 2-inch [5 cm] planks were inserted on end inside the hoops, around the perimeter, for the initial filling. As the silage poured in and its level rose, more planks were added as needed. It may have been the poor farmer's version of a silo, as a cement one was more expensive and more labour intensive. A foundation was not needed, but the bottom was a poured cement floor. This type of silo was also protected from the elements and fit in well with the main barn, as the top of its roof matched the main barn roof.

21-1

From the outside the barn looks complex. There are many corners and extensions, yet just two buildings have been attached to the main barn: the wooden silo and a straw barn.

The silo and barn were built by Queen's Bush settler Alistair Hughes in the 1870s. The traditional 40-foot by 60-foot [12 m x 18 m] main barn frame was built from hardwood beams, mainly maple, elm, and white ash. It was banked into a northern slope for better winter wind protection, and one set of main barn doors on this side leads to the threshing floor. A wooden-tracked hayfork is still visible inside the centre of the roof, with plenty of room for the fork to be pulled above the girts which connect each of the five bents.

Also unusual is a 30-foot long by 16-foot wide [9 m x 4.8 m] timber-frame structure joining the main barn on the south side. The narrowness of the frame makes one ponder why go to all the trouble to raise a frame that is ultimately, in today's world, too small. Looking at it from the inside reminds me of going through a nineteenth-century covered wooden bridge.

Yet, this straw barn was essential for an expanding farm in the late 1800s. Loose hay took up so much room inside a barn that, if a farmer planned to grow more grain for a larger herd

21-2 *The tall, narrow straw barn offers plenty of height on the warm south side to provide easy access and comfort for livestock.*

of cattle or milk cows, he needed more room to store his crops. The narrowness of this barn makes it easier for a farmer to drop the straw without having to fork it over a distance that a hayfork cannot reach. There is still a metal hay track and car inside this narrow barn. The small size also made for some nice neat pens and stalls in the stable, with a centre aisle for easier feeding, milking cows on one side and pigs on the other. Gutters still show where an 18-cow tie-up was housed.

The present owner, Dave McKinnon, is raising beef cattle on the farm and has turned the stabling into loose housing.

The inside of the main barn is a roof frame based on the queen post system. All the main posts are of equal height and long braces angle at 45 degrees from connecting girts to support the purlins. There is a beautiful symmetry and simplicity to this kind of framing.

A one-room school house is located on an acre corner lot just 400 feet [120 m] from the house and barn. It completes the pioneer scene very well—a barn, a house, and a school—the essential tools needed for a farming family to survive. It's easy to imagine the sound of children playing in the school yard and the smell of wood-smoke coming out of the chimney, as the stove was stoked by teachers and students.

21-3 *The queen-post braces supporting the purlins in the straw barn.*

21-4

21-5

21-6

21-4 *A view inside the structure of the silo. Wooden planks would be placed vertically around the inside of the hoops to hold in the grain.*

21-5 *The main barn, the wooden silo and straw barn all well-kept on this 1870s Kinloss Township farm.*

21-6 *The 14-foot by 14-foot by 28-foot high [4 m x 4 m x 8.5 m] timber-frame silo.*

A view inside the larger Purves barn. The barn was split and widened in the space where the ladder stands. The two added middle posts, on either side of the ladder, run 26 feet [8 m] right up to the purlin plate and roof.

22. A Pioneer Barn

The farms along this short, hilly stretch of gravel road in Bentinck Township have many of the features of a hundred years ago. Stone houses are set well back from the road with farmyards containing two or more century-old barns. Board-covered driving sheds and outbuildings are still intact and the scene gives a feeling of stepping back in time.

When Frank Purves took me for a tour of this road, one late-summer's day, I sensed it was meant to be driven in a buggy because there was so much to see. In one mile I counted four stone houses, all built in the nineteenth century by Charlie Willis, a distant relative of Frank's, and many early barns on mud sills with later stable barns added, standing side by side.

One of the earliest barns built in the former Bentinck Township is on Frank's farm. Following English traditions of barn building, this 32-foot by 60-foot [10 m x 18 m] structure, built in 1851, is sitting on mud sills which are still sound.

Inside, the first ground-level floor is divided into sheep pens and, at one end, a pig pen. The original root cellar was dug along that end, 32 feet [10m] long and 10 feet [3m] wide. There are still two stone walls sticking out of the ground, marking the cellar, where the household fruit and vegetables were stored along with turnips for livestock feed.

22-1 *An air photo of the Purves farm from the early 1980s. The present owners have not changed the farm very much except for renovating the house. (courtesy of Frank Purves)*

The first floor is airy with plenty of windows and a high ceiling. Above is the mow where the hay and straw are kept. The mow itself has high sides, with generous head room where the rafters meet the top plate. The timber-frame structure is the queen post style: three-post bents and thick queen post braces running up to meet the purlin plates that span the length of the barn.

22-2 *These barns and the farmyard, formerly owned by Frank Purves, have a classic pioneer look to them.*

22-3 *This barn, built 150 years ago, is still standing on mud sills. There is a high ceiling on the ground floor where the stables are, while the second floor holds the hay and straw.*

22-4 *Inside the ground-level floor of the smaller barn where the livestock was kept. At the far end, a root cellar was dug and sand added to keep turnips for livestock feed and vegetables for the farm family.*

22-5 *The farm buildings and barnyard are on good level ground with nice spacing. The silo was built in 1911. Note the overhang on the small barn gable where the hay fork emerges to pick up the loose hay and straw from a wagon below.*

22-6 *It's easy to tell by the barn boards where this barn was split in the middle. Another 18 feet [5.5 m] were added making the overall width of the barn 54 feet [16.5 m].*

The layout of these buildings is satisfying. Even though it's a very hilly area, with the Allan Park Conservation Area and cross-country skiing one mile east, the farm buildings, house and barn yard are on level ground with roomy spaces between them.

In 1844 the scenery that met four Willis brothers must have been very different. Each took out 50 acres [20 hectares] from the Crown but William must have been the luckiest, since on either side of his farm the land was either hilly or swampy. Huge hardwoods and hemlocks were cleared to make fields for cultivation; then in 1852 there was a tragedy. While cutting down a tree, William Willis was struck and killed by the tree and his son, Henry, took over the farm. William's was the first gravestone at St. George's Cemetery, just east of Hanover, on Grey County Road 4.

The larger barn we see today was built by Henry. At the time it was only 36 feet by 60 feet [11 m x 18 m], not much larger than the original barn, although the newer one had a foundation and stabling below for livestock.

Mark Willis took the farm over in 1905 upon his father's death. At some time in the 1920s he decided to expand the larger barn. A common practice was to split the barn in two and add a section in the middle. This Mark did, expanding the middle by 18 feet [5.5 m] and adding another 10 feet [3 m] to the overall height of the timber frame thus providing more room for loose hay and straw. The original end posts were kept and two 26-foot [8 m] middle posts were added, changing it to a four-post bent from the original three. The fact that there were still such large trees available for the 10-inch by 10-inch [25 cm x 25 cm] maple centre posts speaks of the woods still accessible in the 1920s. It's a magnificent structure, with the middle posts soaring all the way to the roof and purlin beams. The granary was also expanded and a new one built onto the end of the original granary giving a total capacity of more than 45 tons [41,000 kg]. Another 60-foot [18 m] stone foundation was added on the south side to accommodate the split; parts of the old wall are still visible. All in all, a large project that was tackled with skill and stamina, for it was all hand labour accomplished with country expertise.

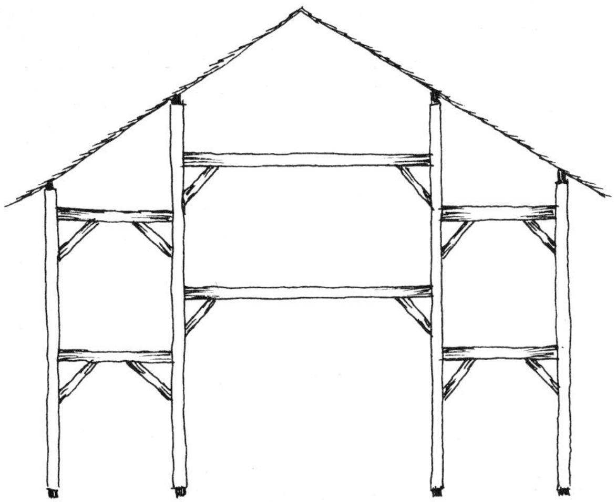

Illustration 1

What many successful farmers wanted by the early twentieth century was a silo. A 12-foot by 30-foot [3.6 m x 9 m] cement structure was commissioned. Still inscribed on the original walls of the silo are the names of those who built it: W. McNaughton, R. Inglis, G. Smith, and the date, June 6, 1911.

Frank is proud of the fact that this silo was larger than most built at the time, and therefore held more silage, which could feed more cattle.

Both hay cars and forks are still visible in the two barns; Frank used them until health reasons forced him to sell the farm in 1991. A nice little overhang on the smaller barn at the top of the gable protects the oak hay car track from the weather. The hay would be brought by wagon in front of the smaller barn, picked up outside by the hayfork, lifted, and brought in through the large upper doors.

Frank was born in the 1920s and has memories of very different times. He recalled that between Hanover and Durham there was a small village every one-and-a-quarter miles [2 km] and at the centre of every village was the hotel or inn. Travelling by buggy or wagon made slow progress and a rest in one of the hotels was almost "mandatory," said Frank. Even though there was no liquor allowed during Prohibition, "it seemed," Frank said, "there was whiskey available in every hotel during the thirties, for 25 cents a gallon!"

Frank recalls seeing the local MPP, Agnes Macphail, of the United Farmers of Ontario. She was born near Ceylon, just west of Flesherton, and Frank remembers her officially opening the Hanover Fair one year. "She looked like a queen," he noted.

In 1943, Walter, the last Willis, took over the farm after his father's death. A shed was built connecting the two barns that were 15 feet [4.6 m] apart. Frank remembers there were three, three-horse teams on this farm and it wasn't until Frank took over in 1967, that the first tractor was bought—a 1959 Ferguson.

"The old times were good. We had a little bit of everything. Common trouble kept people together, that's how they got along. If there was a disaster, they would all help each other," Frank said.

22-7 *The silo built in 1911 with the builders' names forever engraved into the cement. At the top are the owner's initials—MW [Mark Willis], 1911.*

22-8 *The old and the new—Frank Purves, the former owner, and Veronica Verweij, daughter of the new owners—in front of the small shed built in the 1940s to connect the two barns.*

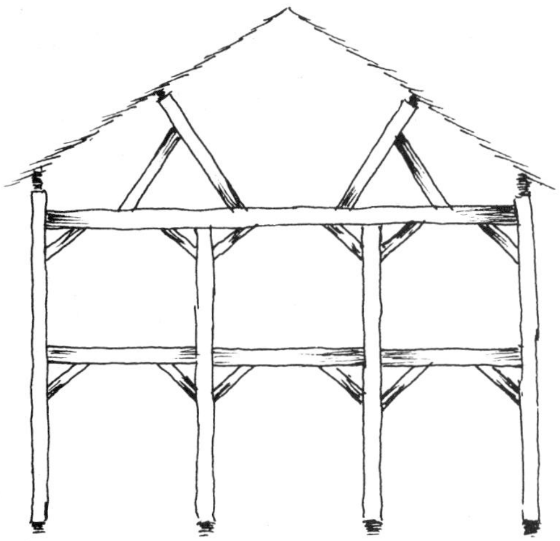

Illustration 2

23. Stonyground Gardens and Farm

Walking around this farm you will come across old implements imbedded in concrete, stone, or wood. Between the house and barn, pathways wind through a huge variety of flowers and vegetables, orchards, and by a pond. A Greek sundial almost blocks one of the walkways. Inside the barn you might find an art show, a poetry reading or live music.

This is Stonyground Gardens, where author, professor, and steward of the farm, Douglas Chambers presents a beautiful variety of nature and art for the public to view, with some of the proceeds going to local charities.

Douglas is a descendant of the original family who took the farm out of the Crown in 1849. Scottish-born Archibald Campbell and his two sons walked to Walkerton and picked out three 50-acre [20 hectare] lots two miles west of the town on the first concession north of the Durham Road in Brant Township, and began a long and mostly prosperous life on the farm.

But there were hardships. Four years after the farms were settled, Archibald's son, Angus, travelled all the way to Buffalo, to bring medicine for his younger brother, Duncan, who was very ill. No one thought young Angus would return, for in those days, a trip to Buffalo would involve trekking for days through wilderness. Angus did come back and the medicine cured Duncan, who lived another 60 years.

More bad luck struck in 1871. Roofers were hammering cedar shingles onto Duncan's new house, intended to replace the log shanty the family was living in. A hot ember came floating through the air from the farm across the road. The farmer was burning a swamp he was trying to clear, but had failed to take into account a strong wind. The burning ember landed on a straw stack next to the first barn built in the 1850s. By the time the roofers came down it was too late to save the barn, which burned to the ground.

23-1 *This barn is surrounded by English-style hedges, banks of flowers and pathways. The original barn, which burned in 1871, was replaced by the present structure. It was expanded by 20 feet [6 m] in 1917 by splitting the barn in half and adding another section to the middle.*

23-2 *Barn raisings brought out the whole neighbourhood. This scene was repeated after a barn fire in 1871 and when the barn was expanded in 1917.(courtesy of County of Grey-Owen Sound Museum)*

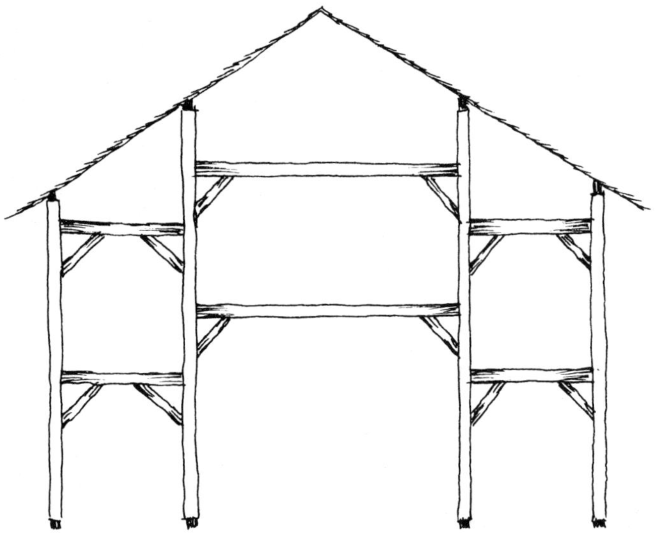

A new 40-foot by 60-foot [12 m x 18 m] traditional timber-frame barn was built that year, with the help of neighbours and the community. Neighbourly help was the insurance policy of the day.

Running a traditional mixed-farm operation of the time, the family raised most of their food from the gardens, orchards, and livestock. Any excess was sold for cash. "Farming has become more and more narrow in scope," comments Douglas. He remembers days when fields were between 5 and 15 acres (2 hectares and 6 hectares] and a mixed operation involved the whole family and the community in threshing and haying bees. Douglas recalls when he was a child, waking up at 5 o'clock to help cook porridge, eggs, ham, biscuits, and pies, and that was just for breakfast!

"There was an enormous friendly rivalry between women about who baked the best," explained Douglas. "People were always at each other's farms to help, to borrow, to chew the fat, and of course, were invited to eat. Farming was a very communal effort back then."

Douglas still bears the scar of landing on a belt powered by a tractor pulley while grain was being threshed in the barn. "I was working on the machine with a threshing gang at our farm and I fell on the belt while forking straw in the mow," said Douglas.

In 1910 another 50 acres [20 hectares] to the northwest was added to the original 100 acres [40 hectares], from which 50 acres had been sold in 1865. An expansion of the barn to house the extra crops was deemed necessary.

It's wonderful to see how imaginative and courageous these early farmers were. In 1917, instead of building a simple addition, the Campbells decided to split the barn in half and make it wider and taller. The east half of the barn was moved 20 feet [6 m] and set on top of wooden pillars. The new timber frame, with four central posts, was inserted to connect the two halves. The expansion is still seen plainly inside the barn frame. The new timbers are all sawn, with location markings still on them, while the rest of the original frame has the rough marks of broadaxe and adze. After the barn was moved the original stone foundation sat 20 feet [6 m] inside the barn; parts of it are still there.

During the 1920s, this was one of the first farms to have pressurized water in the house and barn by using a gasoline-powered motor to drive a pump.

23-3 *The centre of the garden is a flourishing exuberant bank of Brown-Eyed Susans spilling onto the pathway.*

23-4 *When Douglas Chambers inherited the farm he decided to use some of the old farm implements, in this case wood-framed harrows, as part of his Stonyground Gardens.*

Douglas talks of an uncle who farmed the land into the 1970s, still using a one-furrow plow and horses. After his death, a cousin next door bulldozed the rock fences and expanded the fields for modern cultivation. The barn was not being used; at the back of it old implements and farm machinery lay abandoned and rusting in the tall grasses and burdock. When Douglas inherited the farm from his aunt, he decided to do something different.

In 1984 Douglas took a trip to England with two landscape designers to tour and study gardens of the past. "I wanted to see how I could incorporate the traditional English garden into my farm," he said. Out of this tour was born Stonyground Gardens.

At the bottom of a slope, just west of the brick house, was an old gravel pit. Douglas converted it into a pond and surrounded it with water plants, flowers and trees. Walking the many paths extending from the pond, visitors encounter flowers, vegetables, fruit trees, grape vines, and statues. This is not the prim and proper landscape of the English upper crust. Odd and wonderful objects appear at every turn. There is a cultivated wildness to the gardens, where groups of flowers flop over the gravelly paths, Latin-inscribed bronze and marble statues pop up in the middle of paths, and trees cut at the 8-foot [2.4 m] mark create

23-5

23-6

23-5 *Long walkways bordered with columns of planted cedars help form part of Douglas Chambers' interpretation of a traditional English garden.*

23-6 *A Hercules plow stands as a testament to Chambers' forefathers.*

privacy borders. Even within the cropped fields (leased to his cousin) are wide bands of grassways between crops, with parts of the implements Douglas found behind the barn formed into metal art. A two-furrow Hercules plow stands mounted at the end of one such 100-metre walkway, displayed on a concrete base as a testament to Douglas's forefathers.

In 1990, Douglas decided to fix the neglected barn. The predominately white ash and elm structure was not being used, and the west foundation at the gangway was being pushed in by the earth mounded against it. With a bulldozer, parts of the old inner east foundation were pushed against the west side to reinforce it. The foundation posts on the east, added in 1910 with the barn's expansion, were replaced with metal, and new doors were hung to allow a tractor to move inside the stables. All the stalls were removed and now the whole downstairs is used for storage.

With the addition in the mow of a large theatre curtain beside the granary and straw bales stacked as seating, the barn was ready for something the early settlers would never have imagined: art exhibits, concerts, poetry readings, occasionally accompanied by gourmet dinners. "People come here for the different events because of the aura the barn and gardens give off," Douglas said.

Though the farm today is a conventional cash crop operation, the three acres around the house and barn have drawn people from all over Ontario. Stonyground Gardens offers a glimpse of a different age, a slower-paced time where beauty was achieved through unhurried, deliberate work.

23-7 *There is a peaceful atmosphere inside the mow where the stage is set for performance art.*

23-8 *Simple seating in the form of straw bales provides the audience with comfort and some identification with the farm.*

24. The Mill on the Teeswater River

James Stark was an educated man with a vision of a flour-milling enterprise in the village of Paisley on the Teeswater River. "I always aimed at milling," wrote James Stark to his brother, Dan, in Glasgow in 1888. "I have a fine mill with a capacity of 100 barrels of flour in 24 hours. Milling may be said to have reached perfection on this continent—Minneapolis (USA) alone turns out 150,000 to 200,000 barrels per week," added James.

24-1 *Stark's Mill in 1900, taken during a flood in the spring. The mill is the tall white building, with the elevator behind it. The sawmill is on the right with a small bridge leading to it. The 120-foot [36 m] dam at the extreme right corner of the photograph controlled the flow of water to the sawmill and the flour mill. The picture was taken from Stark Vale Cemetery, where the builder, James Stark Sr., was buried. (courtesy of Bruce County Museum & Archives)*

Just before the Teeswater River meets the Saugeen in Paisley a flour mill was built in 1854. In 1873 James Stark, originally from Scotland, bought the mill from John Valentine and began the business of flour milling which was to continue in the same family until 1975.

From some of James Stark's business notebooks, saved by his great-granddaughter Shirley (Stark) Chapuis, we learn of successes and tragedies as they unfolded. "In June 1884 the flour mill with 10,000 barrels wheat was all destroyed by fire, was pretty well insured but sustained a loss of $6,000 to $8,000, after which I rebuilt the present mill."

This is the proud timber-frame structure we see today. Built by barn framers, it stands five storeys high with a four-storey grain elevator next to it. A few years after the great fire James Stark wrote to a friend in Glascow with more bad news. "In December of the same year our oldest son, John, who was the baby in Glasgow, took sick and died after three weeks illness of Bright's disease of the kidneys, leaving a widow and two little boys. She has recently married again (1888). Such is life."

Life did continue and James' only other son, James Jr., sixteen at the time, began his career at the mill.

The mill was rebuilt to the highest standards of the time which even today are impressive. There were two water-powered turbines in the cavernous basement rated together at

24-3 *Five storeys high, the building was rebuilt in 1884 after a fire.*

24-2 *Some of the large gears and massive posts in the cavernous basement of the mill where the 125-horsepower turbines powered the mill.*

125 horsepower. They ran the whole operation. The structure was five storeys of sawn timbers. Each 40-foot by 48-foot [12 m x 14.6 m] floor was supported by five bents with four posts per bent, a total of 20 posts per floor. Wood was not a problem for all of it came from Stark's sawmill, located just upstream from the mill. Each floor had 2-inch [5 cm] thick tongue and groove maple flooring; the walls were all tongue and grooved pine. The floor joists, which had to support the great loads, were 3-inch by 10-inch [7.6 cm x 25 cm] spruce on 20-inch [50 cm] centres. All 12-inch by 12-inch [30 cm x 30 cm] corner posts were double braced; the middle posts had added braces from the post down to the floors. This was a timber-frame structure meant for hard work. (See photo 7)

And work it did. For months on end the mill would run 24 hours a day six, or sometimes seven days a week. Stark would often start at midnight Sunday and grind day and night until midnight Saturday. Five or six men operated the mill during a shift.

24-4

24-5

24-4 *The grain elevator at the right of the mill building could hold up to 640 tons [581,000 kg] of grain. The grain could be augered into the mill via pipes that can be seen under the catwalk between the two buildings.*

24-5 *Every storey of the building still has floor-to-ceiling tongue and groove pine bins into which the grain was augered from the elevator.*

There was a milling-in-transit deal with the railway. For every carload of wheat that came to Paisley, a carload of flour was shipped out. The trains brought in hard, spring wheat from Manitoba and Saskatchewan, via Owen Sound by way of Palmerston, to the small Paisley station. The Starks never had to reject wheat because of a lack of storage capacity. The elevator at the mill could hold 640 tons [581,000 kg] of grain in 15 bins. The bins were made from solid 2-inch x 6-inch [5 cm x 15 cm] lumber stacked and nailed flat and criss-crossed at the ends. The amount of lumber this would take is astounding, but no doubt the lumber came from Stark's own sawmill. Another elevator, constructed at the railway station at the height of the milling operation, provided the same capacity as that of the mill. These structures were built with permanence in mind even though such extravagant use of wood contributed to the devastation of the forests which covered this part of Ontario.

The flour was a blended variety shipped all over the world under brand names such as *Prairie Rose, Pure Cream* and *Emerald Queen*. These blends combined locally grown soft wheat with a mixture of western wheat. The result was a popular product better than any that could be produced without blending. Shirley remembers her great-grandmother was renowned for her delicious golden bread.

But trouble was brewing on the flour-business front. James Stark wrote in 1889 to his fellow Ontario mill business owners, "The yankee miller with cheap wheat, lower rates of freight and trifling duty to pay on flour undersells the whole of us and realizes a profit."

James Stark was a leading member of the Ontario Milling Association which lobbied the federal government to increase the duty paid on US wheat coming into Canada from fifty cents to one dollar per barrel. "I really think Sir John [A. McDonald] cannot change the duty on flour until one-half of us cease to exist, then the remedy will come when the young and aspiring with a little money may step in our shoes and enrich themselves by our loss and misfortunes," wrote an angry James.

With prophetic concern James wrote again a few months later about Canada's policy of trade with the US. "I do not wish to enter the field of politics but often wonder at the remarks of some of our leading papers and

24-6 *From the fourth floor of the mill a door and narrow gangway lead to the top of the grain elevator.*

distinguished men who always denounce the National Policy and who advocate free trade without providing for an equivalent in return. We have the right to demand from our government the same amount of protection in our favour." Almost 110 years later we see the same problems in Canada of free trade coming and threatening to hurt small businesses. The lack of concern by Sir John A's government for the milling industry in Ontario and Quebec spelled its doom and Stark's along with it. In 1917 the mill in Paisley closed its flour-milling production due to the huge amount of flour being produced in the US without an increase in duty.

James Jr. took over the business in 1904 upon his father's death and ran the mill until his death in 1950, when the mill was handed over to his son Nelson.

24-7 *Each floor was held up by 20 beautifully finished posts. The holes in the floor of the third storey were openings for gears and belts that once ran the milling machinery.*

"Life revolved around the mill," said Shirley, a daughter of Nelson. The family homestead was a big rambling house built by James Sr. across the road from the mill and close beside the Teeswater River. Spring break-up of ice and floods were a constant worry for the Starks. The dam had to be always monitored to keep the flow even to the mill and also prevent flooding. The 120-foot [37 m] dam had four slides or openings where 8-inch [20 cm] boards could be pushed down or pulled up to control the amount of water that flowed to the mill. In late winter the Starks had to be especially vigilant. It was a dangerous job, sometimes in the dead of night, to pull up boards submerged under water with a pike pole while standing on a narrow walkway above the rushing river. Floods did come. "The townsfolk used to come and help sandbag and work furiously for days and it wasn't for the money. They just did it to do a good turn for someone," Shirley said.

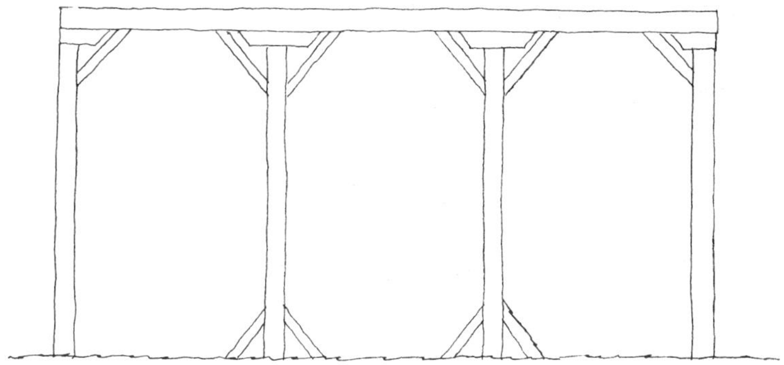

When Shirley was a young girl during the 1940s she used to get paid for sewing the feed sacks. She also remembers the hungry thirties, when flour sacks were used as material for clothing. The family homestead was a farm with pigs raised in the barn on chop produced at the mill. They also grew much of their own food and even though times were hard, there was always plenty of food for the family of four girls, four boys, their father, Nelson, and mother, Irene.

24-8

24-9

24-8 *Each floor at one time was filled with flour- and feed-milling machinery. Here are some seed cleaners. Most of the machinery was sold to local Mennonites when the mill closed in 1975.*

24-9 *Inside the large flour-sifting machines were hand-made wooden auger paddles that augered the flour from one machine to another to produce a fine sifted flour.*

"Entertainment was what you made of it." Shirley remembers riding her bicycle around the empty top floor of the mill, exploring along the water's edge and swimming in the mill pond. "We had a wonderful childhood." On Sundays, townspeople dressed in their finery would picnic along the river's edge and take rowboats out in the mill pond.

Unfortunately, changes in technology were marking the end of the mill's usefulness. By the 1960s most of the business was based on grinding feed from grain either brought in by farmers or shipped by boat into the port of Owen Sound and trucked to Paisley. "I can recall the Little Beaver grain grinders spelled the end of the mill," Shirley said with some frustration. These small grain grinders were marketed for a reasonable price to farmers by the 1960s and although the mill continued, it was with ever-decreasing business, finally closing its doors by 1975, after 124 years of operation.

The mill and property have since been sold but sit unused. The raceway, where the once 12-foot [3.6 m] deep flume flowed, is a mere trickle, dammed up by beavers and filled in with soil.

It is a testament to the forward thinking of James Sr. that the mill property is filled today with large, straight black walnut trees, which he planted 100 years ago for a future highly-valued lumber supply for his children and grandchildren. In front of the Stark homestead more black walnuts spiral into the sky, along with a great oak tree in the front yard.

The Stark Vale cemetery, on land donated by James Sr., holds all the Stark family up to the present day. Beneath a tall tombstone high above the Teeswater River, the mill pond, and the mill itself, lie the remains of James Sr., his wife Margaret, their sons John and James, and Sarah, the wife of James Jr. Their earthly remains overlook everything they lived for.

24-10 *James Stark, Sr., planted black walnut trees 100 years ago around the mill and today those great trees have re-seeded to start new growth of the valuable wood.*

25. The Beginning and the End

25-1 *The beginning: friends and neighbours celebrating the raising of a barn. This was the third barn built on this foundation, after lightning struck the previous two. The front of the foundation is still the old stone, while the back is built from cement blocks made locally. (photograph by J. James, courtesy of County of Grey-Owen Sound Museum)*

The last barn raising on this farm was a momentous event. In 1910 more than 200 friends and neighbours came to help rebuild after a fire gutted the previous barn.

In fact, the Dudgeon farm had suffered some exceptionally bad luck. Their original barn, built in 1880, was struck by lightning in 1910; a new one was rebuilt with the community's help. Just as the last board was nailed onto the siding and the family was sitting down to dinner, lightning again struck the barn and the structure burned to the ground. And they say lightning doesn't strike twice.

For the barn raising in the photograph, the second one that year, the community came out in even greater force because of the Dudgeons' double misfortune.

This was a prosperous mixed 200-acre [80 hectare] farm at the time. Although Holland Township is a municipality of hills and gravel, this farm was fortunate to have some good grain-growing fields to the north of the barn.

Seventy-one years later, after the farm was sold, the large barn was set on fire in the middle of summer by an unknown arsonist who was never caught—the third time the cry of Fire! was heard on this farm. Fortunately the barn contained no livestock and hardly any hay. Now all that remain are parts of the foundation and a few scarred beams.

This barn site, located just north of Chatsworth on Highway 6 & 10, on the Holland Township side, has an interesting foundation that was made locally. When the barn was rebuilt, the original stone foundation was enlarged to accommodate a larger barn frame. The new foundation was made from 60-pound [27 kg] cement blocks manufactured in

Chatsworth. The stone foundation has long since collapsed, but the cement blocks are still there and appear sound. The same blocks form the front of Bill's barbershop in Chatsworth, on the Highway 6 side of town.

After 100 years in use as a farm, the house remains but the barns are passing into memory.

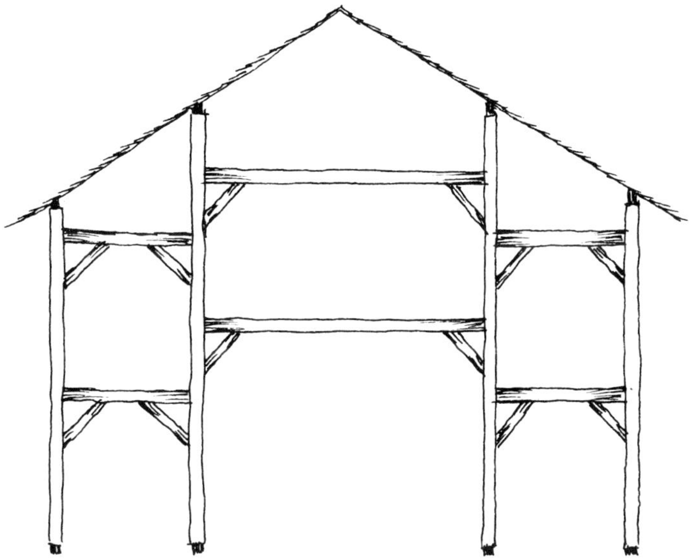

25-2

25-3

25-2 *The end: a spectacular blaze engulfs this barn, just north of Chatsworth, in 1981. Flames at both ends of the barn signal a deliberately set fire. (courtesy of Glen Stephens)*

25-3 *All that's left are cement blocks manufactured in Chatsworth in 1910.*

Types of Wood

BEECH: This wood today is mainly used in cheaper furniture and flooring. A very strong wood, beech was comparable to maple for use in timber framing.

BLACK WALNUT: The wood of the black walnut has excellent technical qualities making it one of the most desirable woods for fine furniture and cabinet making. There are stories of entire barn frames built from this wood.

CEDAR: This wood was most commonly used for fencing, and because of its resistance to rot, was in great demand for barn sidings as well. In barn frames, it was commonly used for braces. Occasionally, a large cedar post supported a cross girt inside the granary of a barn. A light wood, it was used for such household purposes as cedar chests and "moth proof" closets. Coffins were very often made from cedar.

ELM: This tree could be called the "rock" of timber framing. The wood was so hard and able to withstand moisture that it was used predominantly as sill-plate timbers and for other exposed areas of the barn. These majestic hardwoods, which have died out in Ontario due to the Dutch Elm disease, were huge trees that could easily make beams 60 feet [18 m] or longer. As sill plates, where the barn frame met the stone foundation or just the ground, they have lasted the rigours of rain, snow, cold, and heat. In some barns, where elms were plentiful on the homesteads, elm was also used as posts or top plates, where the weather met the rafters.

HEMLOCK: This softwood is very common to Southern Ontario, usually found on heights of land. It can grow to 60 to 70 feet [18 m to 21 m] and have a trunk diameter of 2 to 4 feet [61 cm to 122 cm]. One unfavourable quality of this wood is called "the shakes", a separation of the grain caused by hemlocks' bending in strong winds. Unfortunately, one could not know if the tree had shakes until it was sawn. Hemlock was inexpensive lumber, used extensively in barn building, most frequently for siding.

HICKORY: There were good pockets of hickory in the Queen's Bush, where the soil was best and drainage good. Unfortunately, much of that land was cleared for farmland, removing the hickory. Where all the barn floor joists came together, a strong beam was needed because of the great burden that it would take: this is where the hickory was put to good use. The "summer beam", as it is now called, derived its name from *sumpter*, or burden horse.

MAPLE: The most predominant hardwood in the Queen's Bush, maple was used extensively in barn frames. Readily available and tall—many barn timbers measured 60 feet [18 m] long—maple was not suitable for uses that would place it in contact with moisture. Because of its heaviness, most maple beams were used as upright posts within the bent itself, and not as lateral girts, which would have to be lifted singly.

OAK: Oak was the heaviest wood, and although relatively rare in this area, was chosen in some instances for framing barns and houses. Many of these trees found their way into use as barrel staves because of oak's easy bendability and absence of odour.

SPRUCE: Light, yet the strongest of the softwoods, spruce was very often used in timber barns as posts and girts with medium spans. Many early bridges were made of spruce, as were house frames; the latter practice continues today.

WHITE ASH: The ash tree has many of the qualities of oak. It is a fast, straight-growing tree, and many barns were framed entirely out of white ash although the braces were cedar and the pins were mainly oak. The ash tree could also be used as a cash crop for the early settlers who sold it for splint (baskets and chair bottoms), barrel-hoops, and furniture wood.

WHITE PINE: An extremely light versatile wood, pine does not rot easily, even if wet, and made its way to fill the farmers' needs. Pine was often used as inside barn posts, although there was the occasional barn whose frame was made entirely of pine. (See Barn 3.) Because pine was a lighter wood, timbers used for barns were larger, some as massive as one foot square and forty feet long [30 cm x 12 m]. Most commonly, pine was used as boards for siding a barn, although cedar was considered a better choice.

Barn Locations

Barn 1: **Many Barns in the Making**
Lot 3 Concession 5, Sullivan Township, Grey County

Barn 2: **The Moveable Barns**
Lot 19, Concession 12, St. Vincent Township, Grey County

Barn 3: **The Pine-Framed Barn**
Lot 10, Concession 5 NDR, Bentinck Township, Grey County

Barn 4: **Log and Timber-Framed Barn**
Lot 15, Concession 4 NDR, Glenelg Township, Grey County

Barn 5: **First Quota Chicken Barn**
Lot 13, Concession 9, Sydenham Township, Grey County

Barn 6: **The U-Shaped Barn**
Lot 2, Concession 8, Derby Township, Grey County

Barn 7: **A Sullivan Hardwood Barn**
Lot 26, Concession 13, Sullivan Township, Grey County

Barn 8: **The Biggest Barn in Grey County**
Lots 29 & 30, Concession 2, Normanby Township, Grey County

Barn 9: **Double Barn**
Lots 31 & 32, Concession 12, Brant Township, Bruce County

Barn 10: **Salt Box Barn**
Lot 9, Concession 2, Derby Township, Grey County

Barn 11: **McClure's Mill**
Mill Road, Chesley, Bruce County

Barn 12: **Beaver Bill's Wood Foundation Barn**
Part Lot 29, Concession 2, Elderslie Township, Bruce County

Barn 13: **Four Barns Later**
Lot 2, Concession 15, Proton Township, Grey County

Barn 14: **Royston Park: A Gentleman's Farm**
Lots 10 & 11, Concession 3, Sarawak Township, Grey County

Barn 15: **The Stone and Timber-Framed Barn**
Lot 22, Concession 1, Bentinck Township, Grey County

Barn 16: **Orchard Barns**
Lot 31, Concession 11, Collingwood Township, Grey County

Barn 17: **Twin Barns**
Part Lot 8, Eastern Division, Ashfield Township, Huron County

Barn 18: **The Leaning Barn of Crawford**
Lot 26, Concession 10 NDR, Bentinck Township, Grey County

Barn 19: **A Different Barn Roof-Line**
Lot 21, Concession 13, Sullivan Township, Grey County

Barn 20: **The Symphony Hall Barn**
Lots 43 & 44, Concession 3 EGR, Glenelg Township, Grey County

Barn 21: **Timber-Frame Silo and Barn**
Lot 11, Concession 5, Kinloss Township, Bruce County

Barn 22: **A Pioneer Barn**
Lot 35-38, Concession 2 SDR, Bentinck Township, Grey County

Barn 23: **Stonyground Gardens and Farm**
Lots 8 & 9, Concession 1 NDR, Brant Township, Bruce County

Barn 24: **The Mill on the Teeswater River**
Part Lot 51, Concession 22, Greenock Township, Bruce County

Barn 25: **The Beginning and the End**
Lot 1, Concession 1E, Holland Township, Grey County

Sources

Map of the Queen's Bush from *The Home Knowledge Atlas*, Home Knowledge Association, 1889.

A History of Normanby Township, Campbell Cork, editor, The Township of Normanby, 1989.

Memoir of Mrs. Guse, *Daily Sun-Times,* Owen Sound, 1933.

Proud People, The Lindsay Township History Book, Allan Bartley, editor, Lindsay Township Historical Society, 1987.

About the Author

Jon Radojkovic, known to his neighbours as Jon Log, writes, builds, and farms in Sullivan Township, northeast of Chesley, Ontario.

In the last ten years Jon has freelanced as a photographer and journalist with the Owen Sound daily newspaper, *The Sun Times,* and local weeklies such as *The Chesley Enterprise* and *The Markdale Standard.* Numerous articles and poems written by Jon have appeared in *Mosaic*, an arts and lifestyle magazine originating in Durham.

(courtesy of Dyan Jones)

Jon's interest in historic buildings began when he attended the University of Toronto where he received his Bachelor of Arts. As part of his course, he studied the identity and origins of rural and small-town Ontario architectural design.

Since the late 1980s, Jon has been constructing timber-frame homes and farm structures, beginning with his own house in 1987. Some of his buildings, including his own home, incorporate timbers from barns he dismantled. "Old barn frames fascinate me," he says. "When I look at them, I want to know the stories of the people who built and lived with those massive barns."

Jon, his wife, Lillian Burgess, and three sons grow garlic, asparagus and shiitake mushrooms, as well as some grain crops on their 96-acre (38.5 ha) farm. They have also planted a nut and fruit orchard and more than 10,000 other trees on the rougher parts of their land.

About the Artist

An artist who looks for "remnants of history peeking tentatively from forest floor [or] ploughed field", Mary Tripp MacCarl lives in a log cabin beside the Rocky Saugeen River near Durham, Ontario. It's her home and her Stones and Cedar Studio, where she creates works using pencil, clay, and found objects.

Mary began drawing as a child, and later trained at the New School of Art in Toronto, and at the Ontario College of Art. She has shared her artistic intuition with students at Georgian College, Pinehill Youth Residence, and in private classes.

Her works have been exhibited at the Tom Thomson Memorial Art Gallery, and the Durham Art Gallery, as well as to the many visitors who follow the Autumn Leaves Studio Tour each October.

Mary says, "I grew up on a dairy farm in the Niagara region, and have always loved barns. Our barn became a play place... a work place... and a discovery place. In checking resources for the drawings in Jon's book, I found I was relating much of the information to the barn I remember so well."